W9-CBO-268

[Eroded Rock]

BANFF NATIONAL PARK, ALBERTA, CANADA

Banff National Park is located in the Canadian Rocky Mountains in the province of Alberta. Banff National Park covers 6,641 square kilometers (2,564 sq. mi.) and includes breathtaking mountains, dense forests, and numerous glaciers.

In the park, you can find many types of sedimentary rocks, including sandstone and limestone. Over time, much of this rock has been eroded away. The large rock in the picture was once part of a mountain. The river eroded the mountain and now the rock stands alone.

Banff National Park was named a UNESCO World Heritage Site because of its mountain peaks, glaciers, lakes, waterfalls, canyons, and caves, as well as its many fossils. Conservation and preservation are priorities in this park that has so many natural wonders.

NATIONAL GEOGRAPHIC
SCIENCE

EARTH SCIENCE

NATIONAL GEOGRAPHIC
School Publishing

PROGRAM AUTHORS

Kathy Cabe Trundle, Ph.D.

Randy Bell, Ph.D.

Malcolm B. Butler, Ph.D.

Judith S. Lederman, Ph.D.

David W. Moore, Ph.D.

Program Authors

KATHY CABE TRUNDLE, PH.D.

Associate Professor of Early Childhood Science Education, The School of Teaching and Learning, The Ohio State University, Columbus, Ohio
SCIENCE

RANDY BELL, PH.D.

Associate Professor of Science Education, University of Virginia, Charlottesville, Virginia
SCIENCE

MALCOLM B. BUTLER, PH.D.

Associate Professor of Science Education, University of South Florida, St. Petersburg, Florida
SCIENCE

JUDITH SWEENEY LEDERMAN, PH.D.

Director of Teacher Education,
Associate Professor of Science Education,
Department of Mathematics and Science Education,
Illinois Institute of Technology, Chicago, Illinois
SCIENCE

DAVID W. MOORE, PH.D.

Professor of Education,
College of Teacher Education and Leadership,
Arizona State University, Tempe, Arizona
LITERACY

Program Reviewers

Amani Abuhabsah
Teacher
Dawes Elementary
Chicago, IL

Maria Aida Alanis, Ph.D.
Elementary Science
Instructional Coordinator
Austin Independent School
District
Austin, TX

Jamillah Bakr
Science Mentor Teacher
Cambridge Public Schools
Cambridge, MA

Gwendolyn Battle-Lavert
Assistant Professor of Education
Indiana Wesleyan University
Marion, IN

Carmen Beadles
Retired Science Instructional
Coach
Dallas Independent School
District
Dallas, TX

Andrea Blake-Garrett, Ed.D.
Science Educational Consultant
Newark, NJ

Lori Bowen
Science Specialist
Fayette County Schools
Lexington, KY

Pamela Breitberg
Lead Science Teacher
Zapata Academy
Chicago, IL

Carol Brueggeman
K–5 Science/Math Resource
Teacher
District 11
Colorado Springs, CO

Program Reviewers continued
on page iv.

Acknowledgments

Grateful acknowledgment is given to
the authors, artists, photographers,
museums, publishers, and agents for
permission to reprint copyrighted
material. Every effort has been made
to secure the appropriate permission.
If any omissions have been made or
if corrections are required, please
contact the Publisher.

Illustrator Credits
All illustrations by Precision Graphics.
All maps by Mapping Specialists.

Photographic Credits
Front Cover Kevin M. Law/Alamy
Images.

Credits continue on page EM10.

The National Geographic Society
John M. Fahey, Jr.,
President & Chief Executive Officer

Gilbert M. Grosvenor,
Chairman of the Board

Copyright © 2011 The Hampton-
Brown Company, Inc., a wholly
owned subsidiary of the National
Geographic Society, publishing
under the imprints National
Geographic School Publishing and
Hampton-Brown.

National Geographic School Publishing
Hampton-Brown
www.myNGconnect.com

Printed in the USA.
RR Donnelley
Jefferson City, MO

ISBN: 978-0-7362-7763-1

11 12 13 14 15 16 17 18 19 20

5 6 7 8 9 10

Miranda Carpenter
Teacher, MS Academy Leader
Imagine School
Bradenton, FL

Samuel Carpenter
Teacher
Coonley Elementary
Chicago, IL

Diane E. Comstock
Science Resource Teacher
Cheyenne Mountain School
District
Colorado Springs, CO

Kelly Culbert
K–5 Science Lab Teacher
Princeton Elementary
Orange County, FL

Karri Dawes
K–5 Science Instructional
Support Teacher
Garland Independent
School District
Garland, TX

Richard Day
Science Curriculum Specialist
Union Public Schools
Tulsa, OK

Michele DeMuro
Teacher/Educational
Consultant
Monroe, NY

Richard Ellenburg
Science Lab Teacher
Camelot Elementary
Orlando, FL

Beth Faulkner
Brevard Public Schools
Elementary Training Cadre,
Science Point of Contact,
Teacher, NBCT
Apollo Elementary
Titusville, FL

Kim Feltre
Science Supervisor
Hillsborough School District
Newark, NJ

Judy Fisher
Elementary Curriculum
Coordinator
Virginia Beach Schools
Virginia Beach, VA

Anne Z. Fleming
Teacher
Coonley Elementary
Chicago, IL

Becky Gill, Ed.D.
Principal/Elementary Science
Coordinator
Hough Street Elementary
Barrington, IL

Rebecca Gorinac
Elementary Curriculum Director
Port Huron Area Schools
Port Huron, MI

Anne Grall Reichel Ed. D.
Educational Leadership/
Curriculum and Instruction
Consultant
Barrington, IL

Mary Haskins, Ph.D.
Professor of Biology
Rockhurst University
Kansas City, MO

Arlene Hayman
Teacher
Paradise Public School District
Las Vegas, NV

DeLene Hoffner
Science Specialist, Science
Methods Professor,
Regis University
Academy 20 School District
Colorado Springs, CO

Cindy Holman
District Science Resource
Teacher
Jefferson County Public
Schools
Louisville, KY

Sarah E. Jesse
Instructional Specialist for
Hands-on Science
Rutherford County Schools
Murfreesboro, TN

Dianne Johnson
Science Curriculum Specialist
Buffalo City School District
Buffalo, NY

Kathleen Jordan
Teacher
Wolf Lake Elementary
Orlando, FL

Renee Kumiega
Teacher
Frontier Central School District
Hamburg, NY

Edel Maeder
K–12 Science Curriculum
Coordinator
Greece Central School District
North Greece, NY

Trish Meegan
Lead Teacher
Coonley Elementary
Chicago, IL

Donna Melpolder
Science Resource Teacher
Chatham County Schools
Chatham, NC

Melissa Mishovsky
Science Lab Teacher
Palmetto Elementary
Orlando, FL

Nancy Moore
Educational Consultant
Port Stanley, Ontario, Canada

Melissa Ray
Teacher
Tyler Run Elementary
Powell, OH

Shelley Reinacher
Science Coach
Auburndale Central Elementary
Auburndale, FL

Kevin J. Richard
Science Education Consultant,
Office of School Improvement
Michigan Department
of Education
Lansing, MI

Cathe Ritz
Teacher
Louis Agassiz Elementary
Cleveland, OH

Rose Sedely
Science Teacher
Eustis Heights Elementary
Eustis, FL

Robert Sotak, Ed.D.
Science Program Director,
Curriculum and Instruction
Everett Public Schools
Everett, WA

Karen Steele
Teacher
Salt Lake City School District
Salt Lake City, UT

Deborah S. Teuscher
Science Coach and
Planetarium Director
Metropolitan School District
of Pike Township
Indianapolis, IN

Michelle Thrift
Science Instructor
Durrance Elementary
Orlando, FL

Cathy Trent
Teacher
Ft. Myers Beach Elementary
Ft. Myers Beach, FL

Jennifer Turner
Teacher
PS 146
New York, NY

Flavia Valente
Teacher
Oak Hammock Elementary
Port St. Lucie, FL

Deborah Vannatter
District Coach, Science
Specialist
Evansville Vanderburgh School
Corporation
Evansville, IN

Katherine White
Science Coordinator
Milton Hershey School
Hershey, PA

Sandy Yellenberg
Science Coordinator
Santa Clara County Office
of Education
Santa Clara, CA

Hillary Zeune de Soto
Science Strategist
Lunt Elementary
Las Vegas, NV

EARTH SCIENCE

CONTENTS

What Is Earth Science?..2

Meet a Scientist 4

CHAPTER 1

How Do Earth and Its Moon Move?..............5

Science Vocabulary..8

Comparing the Earth, Moon, and Sun 10

Earth Spins on Its Axis Science in a Snap! 12

Earth Moves Around the Sun 18

Observing Stars and the Moon 26

The Moon Moves Around Earth 32

NATIONAL GEOGRAPHIC Telling Time Yesterday and Today 36

Conclusion and Review 40

NATIONAL GEOGRAPHIC **EARTH SCIENCE EXPERT:** Cultural Astronomer 42

NATIONAL GEOGRAPHIC **BECOME AN EXPERT:** The Earth-Moon-Sun System:
How Knowledge Grows 44

CHAPTER 2

How Are Rocks Alike and Different?...........53

Science Vocabulary 56

Rocks and Minerals 58

Properties of Minerals 62

Types of Rocks Science in a Snap! 70

NATIONAL GEOGRAPHIC Diamonds and Their Uses 78

Conclusion and Review 80

NATIONAL GEOGRAPHIC **EARTH SCIENCE EXPERT:** Sculptor 82

NATIONAL GEOGRAPHIC **BECOME AN EXPERT:** The Grand Canyon:
History Written in Rock 84

Student eEdition | Vocabulary Games | Digital Library | Enrichment Activities

CHAPTER 3

What Are Renewable and Nonrenewable Resources? .... 93 ⓔ

Science Vocabulary .... 96 ⓟ

Natural Resources .... 98 ⓔ ⊕

Renewable Resources Science in a Snap! .... 100 ⓔ

Nonrenewable Resources 106

Soil .... 114 ⓔ

People and Resources .... 118 ⓔ

NATIONAL GEOGRAPHIC Resources in the Borneo Rainforest 126

Conclusion and Review 128

NATIONAL GEOGRAPHIC **EARTH SCIENCE EXPERT:** Naturalist .... 130 ⓔ

NATIONAL GEOGRAPHIC **BECOME AN EXPERT:** Making Blue Jeans: Resources on the Move .... 132 ⓔ

CHAPTER 4

How Do Slow Processes Change Earth's Surface? .... **141** ⓔ

Science Vocabulary .... 144 ⓐ

Landforms on Earth's Surface 146

Weathering .... 148 ✉

Causes of Weathering *Science in a Snap!* .... 152 ⊕

Erosion and Deposition 158

Causes of Erosion and Deposition .... 160 ✉

Weathering and Erosion Affect People 166

NATIONAL GEOGRAPHIC Shrinking Glaciers in South America .... 172 ✉

Conclusion and Review 174

NATIONAL GEOGRAPHIC **EARTH SCIENCE EXPERT:** Glaciologist .... 176 ✉

NATIONAL GEOGRAPHIC **BECOME AN EXPERT:** Yosemite Valley: Shaped by Weathering and Erosion .... 178 ✉

CHAPTER 5

What Changes Do Volcanoes and Earthquakes Cause? .... **189** ⓔ

Science Vocabulary .... 192 ⓐ

Earth's Structure 194

Earthquakes *Science in a Snap!* .... 196 ⊕

Volcanoes .... 200 ✉

Landslides 204

NATIONAL GEOGRAPHIC Living on the Edge in Japan 206

Conclusion and Review 208

NATIONAL GEOGRAPHIC **EARTH SCIENCE EXPERT:** Volcanologist .... 210 ✉

NATIONAL GEOGRAPHIC **BECOME AN EXPERT:** The Hawaiian Islands: Formed by Volcanoes .... 212 ✉

TECHTREK
myNGconnect.com

Student eEdition

Vocabulary Games

Digital Library

Enrichment Activities

CHAPTER
6

What Can We Observe About Weather? . . . . **221** ⊜

Science Vocabulary 224

Air and the Atmosphere Science in a Snap! 226

Parts of Weather 230

Clouds 238

Air Masses and Weather Fronts 246

NATIONAL GEOGRAPHIC **Climbing Weather** 254

Conclusion and Review 256

NATIONAL GEOGRAPHIC **EARTH SCIENCE EXPERT:** Meteorologist 258

NATIONAL GEOGRAPHIC **BECOME AN EXPERT:** Extreme Weather 260

Glossary EM1

Index EM3

EARTH SCIENCE

What Is Earth Science?

Earth science investigates all aspects of our home planet from its changing surface, to its rocks, minerals, water, and other resources. It also includes the study of Earth's atmosphere, weather, and climates. As Earth is an object in space, Earth science also includes the study of Earth's relationship with the sun, moon, and stars. People who study our planet are called Earth scientists.

You will learn about these aspects of Earth science in this unit:

HOW DO EARTH AND ITS MOON MOVE?

Earth scientists use telescopes and other tools to study Earth and the moon, and how they move. Earth's movement causes day and night and enables you to see different stars at different times of the year. The way the moon moves around Earth causes the moon to look different from Earth at different times.

HOW ARE ROCKS ALIKE AND DIFFERENT?

Rocks can tell Earth scientists a lot about different places on Earth. Earth scientists observe the properties of rocks and minerals and think about how rocks were formed. They study fossils in rocks to learn more about what Earth was like long ago.

WHAT ARE RENEWABLE AND NONRENEWABLE RESOURCES?

Earth's natural resources, such as wind, oil, and metals, are important to people. Some are renewable while others are not. Earth scientists study the natural resources on Earth and how to use them wisely.

HOW DO SLOW PROCESSES CHANGE EARTH'S SURFACE?

Earth Scientists study the slow changes to Earth's surface. Weathering, erosion, and deposition shape and reshape the land. Wind, water, gravity, ice, temperature, and plants are involved in these processes.

WHAT CHANGES DO VOLCANOES AND EARTHQUAKES CAUSE?

Earthquakes, volcanoes, and landslides are natural processes that can change Earth's surface quickly. It is important for Earth scientists to study these processes so they can help people survive these natural events.

WHAT CAN WE OBSERVE ABOUT WEATHER?

Temperature, humidity, wind, air pressure and precipitation are parts of weather. Earth scientists study the weather so they can better predict what weather conditions will occur in the near future.

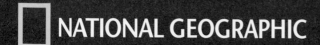
MEET A SCIENTIST

Beverly Goodman: Geo-Archaeologist

Beverly Goodman is a geo-archaeologist and National Geographic Emerging Explorer. As the title "geo-archaeologist" suggests, Beverly blends skills from archeology, geology, and anthropology. She uses skills and knowledge from these scientific studies to explore the complex ways nature and humans interact on coastlines. "Analyzing the causes and effects of ancient environmental events like tsunamis can help tell us which types of coasts are at greatest risk, and what kind of damage to expect in the future," Beverly explains.

Beverly's current focus is in Caesarea, Israel, where Herod the Great built a massive harbor at the end of the 1st Century B.C. Her team's findings prove that a tsunami struck the ancient harbor sometime in the 1st or 2nd Century A.D., and likely caused its destruction.

The tsunami studies that now consume Beverly's schedule began quite by accident. She originally came to Caesarea to help explore the harbor's construction.

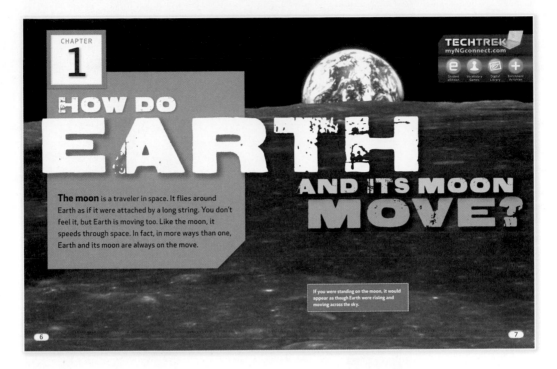

CHAPTER 1

HOW DO EARTH AND ITS MOON MOVE?

The moon is a traveler in space. It flies around Earth as if it were attached by a long string. You don't feel it, but Earth is moving too. Like the moon, it speeds through space. In fact, in more ways than one, Earth and its moon are always on the move.

If you were standing on the moon, it would appear as though Earth were rising and moving across the sky.

TECHTREK
myNGconnect.com

After reading Chapter 1, you will be able to:

- Compare and contrast the characteristics of the sun, moon and Earth.
  COMPARING THE EARTH, MOON, AND SUN

- Explain that the spin of Earth creates day and night. **EARTH SPINS ON ITS AXIS**

- Describe the orbit of the Earth around the sun as it defines a year
  EARTH MOVES AROUND THE SUN

- Explain that Earth's movement and position cause the seasons.
  EARTH MOVES AROUND THE SUN

- Identify the sun, other stars and the moon as common objects in the sky.
  OBSERVING STARS AND THE MOON

- Describe the motion of the moon around Earth. **THE MOON MOVES AROUND EARTH**

- Explain how the visible shape of the moon follows a predictable cycle.
  THE MOON MOVES AROUND EARTH

- Science in a Snap! Explain that the spin of Earth creates day and night. **EARTH SPINS ON ITS AXIS**

HOW DO EAR

The moon is a traveler in space. It flies around Earth as if it were attached by a long string. You don't feel it, but Earth is moving too. Like the moon, it speeds through space. In fact, in more ways than one, Earth and its moon are always on the move.

TECHTREK
myNGconnect.com

Student eEdition

Vocabulary Games

Digital Library

Enrichment Activities

TH AND ITS MOON MOVE?

If you were standing on the moon, it would appear as though Earth were rising and moving across the sky.

SCIENCE VOCABULARY

rotate (RO-tāt)

To **rotate** is to spin around. (p. 12)

As Earth rotates, a place will face the sun and have day and then spin away from the sun and have night.

Europe

South America

axis (AK-sis)

An **axis** is an imaginary line around which Earth spins. (p. 12)

Earth rotates on an axis that runs through its North Pole and South Pole.

revolve (re-VAWLV)

To **revolve** is to move around another object. (p. 18)

It takes one year for Earth to revolve around the sun.

March

June

December

September

my Science Vocabulary

axis (AK-sis)	**phase** (FĀS)
gravity (GRA-vi-tē)	**revolve** (re-VAWLV)
latitude (LA-ti-tūd)	**rotate** (RO-tāt)

TECHTREK
myNGconnect.com

Vocabulary Games

gravity (GRA-vi-tē)

Gravity is a force that pulls objects toward each other. (p. 18)

> Gravity is the force that keeps the moon traveling around Earth.

latitude (LA-ti-tūd)

Latitude is how far north or south of the Equator a place is. (p. 21)

> During summer, the sun does not set at the high latitude of the North Pole.

phase (FĀS)

A **phase** is a lighted part of the moon as it appears from Earth. (p. 35)

> You see different phases of the moon at different times of the month.

Comparing the Earth, Moon, and Sun

What do you see when you look at the sky? If it's a clear night, you will probably see thousands of twinkling stars. You might see the moon in one shape or another. If it's day, the nearest star, the sun, lights up the sky. The sun lies at the center of a system of objects that move around it in space. Earth and its moon are part of this solar system.

The sun is huge compared to Earth and its moon. If the sun were the size of a basketball, look at how small Earth would be. Earth's moon would be the size of the period at the end of this sentence.

← **Size of the sun**

← **Size of Earth**

The moon can be seen in the sky sometimes at night and other times during the day.

Distances between Earth, its moon, and the sun are vast. If you were in a race car traveling at 160 kilometers per hour (about 100 miles per hour), it would take you 100 days to reach the moon. It would take you about 106 *years* to travel to the sun!

SUN

STRUCTURE: Ball of hot gases
SURFACE: Hot gases; no life
DIAMETER: 1,391,980 kilometers (about 856,000 miles)
DISTANCE FROM EARTH: 149,597,892 kilometers (about 93,000,000 miles)
AGE: 4.55 billion years

EARTH

STRUCTURE: Solid and liquid rock
SURFACE: Rock and water; layer of air; has life
DIAMETER: 12,742 kilometers (about 7,900 miles)
AGE: 4.54 billion years

MOON

STRUCTURE: Solid rock
SURFACE: Rock and dust; trace of ice; no air; no life
DIAMETER: 3,476 kilometers (2,160 miles)
DISTANCE FROM EARTH: 384,400 kilometers (about 238,000 miles)
AGE: 4.5 billion years

Before You Move On

1. Name two objects that move around the sun.
2. How is the sun's structure different from the structure of Earth and the moon?
3. **Infer** When you observe the sun and the moon in the sky, they appear to be about the same size. If the sun is so much larger than the moon, why does the moon appear as big as the sun in the sky?

Earth Spins on Its Axis

Think about what happens when you **rotate**, or spin around. As you turn, you face different directions and see different things. Places on Earth are like that. Earth does not stand still. It rotates on an imaginary line called an **axis** that runs through the North and South poles.

Half of Earth always faces the sun and is lit by sunlight. There it is day. The other half of Earth faces away from the sun and is dark. There it is night.

> This photo was taken in the middle of the afternoon in Bolivia, South America.

Europe

South America

Science in a Snap! Night and Day

Mark your location on a globe with a piece of masking tape. Using the flashlight to stand for the sun, shine it on the globe so that it is day where you live.

Rotate the globe so that it is night where you live.

What happened to the place on the globe where it used to be day? How often does this change happen? Why does it happen?

As Earth rotates, the half that had day moves to face dark outer space. The half that had night moves to face the sun. Look at the globe. South America is facing the sun and has day. At the same time, Europe is facing away from the sun and has night. Can you infer what will happen as Earth continues to rotate? South America will turn away from the sun and have night. Europe will turn toward the sun and have day. This cycle of day and night repeats every 24 hours. That's why one complete day lasts 24 hours.

TECHTREK
myNGconnect.com

Digital Library

When it's the middle of the afternoon in Bolivia, what time is it in Greece, Europe?

The Changing Day Sky Suppose you wake very early while it is still dark. Soon the eastern sky begins to brighten. Slowly the sun appears and seems to rise into the sky. By noon, it will be high in the sky. Later in the day, it appears toward the west. There it seems to slowly sink out of the sky. After sunset, the sun is out of view.

Every day the sun appears to move across the sky. But it is Earth that is moving, not the sun.

You can tell about the time of day by observing the position and length of shadows.

early morning

At sunrise, your half of Earth is moving out of darkness and into sunlight. The sun is toward the east. Shadows are long and stretch toward the west. As Earth rotates, your location moves more in line with the sun. The sun is high in the sky, and shadows are shortest. Earth continues to turn, and you move away from the sun. It appears now toward the west. Shadows lengthen and stretch toward the east. The shadows disappear when Earth turns into the darkness of night.

What time of day might it have been when this photo was taken in the Sahara? How do you know?

noon

late afternoon

The Changing Night Sky On many nights you can see the moon in the sky. But it doesn't appear in the same place. Just like the sun, the moon appears to rise toward the east, move across the sky, and set toward the west. The same thing happens with many stars. The same pattern of stars in the sky moves across the sky from the east toward the west. But like the sun, the moon and stars are not really moving this way. They just seem to move across the sky because Earth is rotating.

The moon appears to rise toward the east.

The moon rises higher in the sky as Earth continues to rotate.

Day after day and night after night, the pattern of the sun, moon, and some stars appearing to move across the sky continues. Over time, the patterns repeat so you can predict where these objects will be and how they will appear to move. But remember, these objects only look like they are moving this way because Earth continues to rotate.

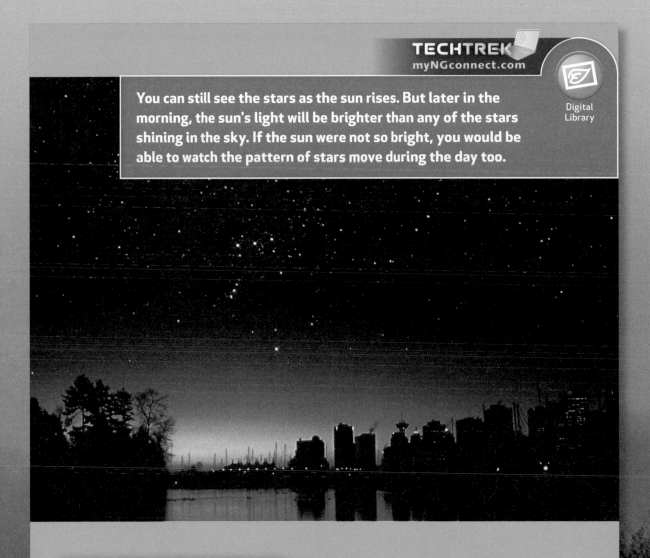

You can still see the stars as the sun rises. But later in the morning, the sun's light will be brighter than any of the stars shining in the sky. If the sun were not so bright, you would be able to watch the pattern of stars move during the day too.

Before You Move On

1. What does *rotate* mean?
2. How does Earth's rotation cause day and night?
3. **Generalize** What can you say about how the sun, moon, and some stars seem to move?

Earth Moves Around the Sun

Rotation is just one way that Earth moves. Earth also journeys through space as it **revolves**, or moves around, the sun. Earth makes one complete revolution around the sun in a year. The path that Earth takes around the sun is its orbit.

Earth stays in orbit around the sun because of **gravity**. Gravity is a force that pulls objects toward each other. The sun's gravity is so strong that it pulls on Earth even though Earth is far away.

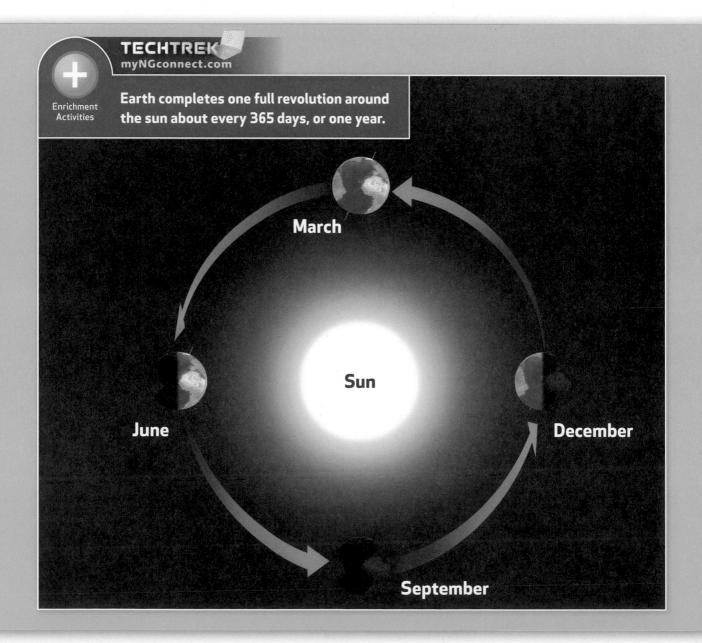

TECHTREK
myNGconnect.com

Enrichment Activities

Earth completes one full revolution around the sun about every 365 days, or one year.

March

Sun

June

December

September

Observe Earth in the diagram. What do you notice about its axis? The axis is tilted, and it keeps this same tilt throughout its orbit. For example in June one part of Earth is tilted toward the sun. At this time the sunlight is less spread out on that part of Earth. Because Earth keeps the same tilt as it moves around the sun, that same part of Earth faces away from the sun in December. The sunlight is more spread out on that part of Earth. The way in which the sun's light strikes Earth at different times of the year causes the changing seasons.

The athlete keeps the heavy ball in a circle by pulling on it. The sun's pulling force, or gravity, works in a similar way to keep Earth in orbit.

Earth's axis is not straight up and down. It is tilted on an angle.

Because Earth is shaped like a ball, sunlight hits some areas more directly than others. Where it hits directly, the light is concentrated and temperatures are very warm. In other places, the light spreads out over a larger area. Temperatures are cooler.

If Earth were not tilted on its axis, the same places would always receive either direct or spread-out sunlight. A place would have about the same temperatures all year long. It would be like one continuous season. But Earth is tilted. So as Earth moves in its orbit, a place receives more concentrated sunlight or more spread out sunlight at different times throughout the year. So the seasons change.

Light that hits the ball near the top is more spread out than the same amount of light that hits near the middle. The same thing happens with sunlight on Earth.

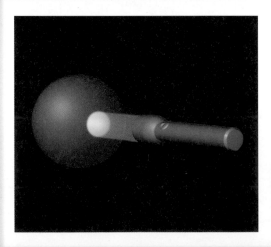
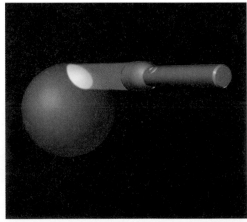

Seasonal changes depend partly on **latitude**. Latitude is how far north or south of the Equator a place is. Places at low latitudes always receive direct or nearly direct sunlight, so temperatures at low latitudes are usually warm. Places at high latitudes never receive direct sunlight. Temperatures there stay cool. Seasons do not change much in either place.

Places in the mid-latitudes, like most of the United States, have bigger changes of seasons. But still, seasons are not all alike. Ohio, for example, has warm summers and snowy winters. In south Texas, however, the temperatures are hot in the summers and you would wear only a light jacket in the winter.

During summer, areas around the North Pole are in sunlight 24 hours each day. This photo was made by taking pictures of the sun at different times during the hours when you would be sleeping! The pictures were then put together.

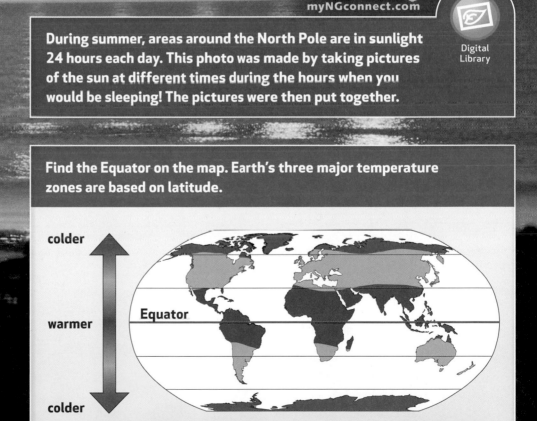

Find the Equator on the map. Earth's three major temperature zones are based on latitude.

colder

warmer

colder

Equator

The Seasons

When the Northern Hemisphere is tilted toward the sun, it's summer there. This hemisphere receives more direct rays from the sun. It also receives more hours of sunlight than at other times of the year. Therefore, this hemisphere is heated most during the summer. Summer temperatures are warm. The sun is higher in the sky, and shadows are short in the middle of the day.

SUMMER Summer in the Northern Hemisphere begins in June. In Reykjavik, Iceland, average high temperatures at this time are 10°C (50°F).

Shadows are shortest in summer.

As Earth continues revolving, the Northern Hemisphere is no longer tilted toward the sun. It receives less direct rays and fewer hours of sunlight than in summer. Therefore, it is heated less. Temperatures are cooler. The sun is lower in the southern sky than in summer, and shadows are longer in the middle of the day. It's fall.

FALL Fall in the Northern Hemisphere begins in September. Cooler temperatures cause the leaves to change colors near Beijing, China.

Shadows are longer in fall than in summer.

Earth continues in its path around the sun. It eventually reaches a point when the northern hemisphere is tilted away from the sun. It's winter there. The hemisphere receives less direct rays from the sun. It also receives fewer hours of sunlight than at other times of the year. Temperatures are coolest in winter. The sun moves lowest in the sky, and shadows are longer in the middle of the day than at any other time.

WINTER Winter in the Northern Hemisphere begins in December. Cold weather brings snow to Winnipeg, Canada.

Shadows are longest in winter.

Earth continues revolving. Winter turns to spring. As in fall, the Northern Hemisphere does not tilt toward or away from the sun. It receives about the same amount of sunlight as in fall. Temperatures increase. They become warmer than in winter but are still cooler than in summer. The sun is higher in the sky than in winter, and shadows grow shorter in the middle of the day.

SPRING Spring in the Northern Hemisphere begins in March. It's time to put away winter coats in Ukraine.

Shadows are longer in spring than in summer.

Before You Move On

1. How long does Earth's revolution around the sun take?
2. How does Earth's tilt help cause the seasons?
3. **Infer** What season is the Southern Hemisphere having when the Northern Hemisphere is having summer? Explain.

Observing Stars and the Moon

Earth's position in space changes as it revolves around the sun. You see different parts of the sky at night depending on the time of year. So in the United States, you might see some stars only in summer and others only in winter. The diagram shows how this happens.

You can see the main stars that make up star patterns such as Lyra and Orion. But the sky contains many millions of stars that you cannot see with just your eyes. They are too dim or too far away.

Lyra and Orion are groups of stars that form patterns. Lyra is seen in June because the night side of Earth faces the part of the sky that contains this star pattern. The day side of Earth faces Orion, so its starlight is lost in the bright sunlight. What happens in December?

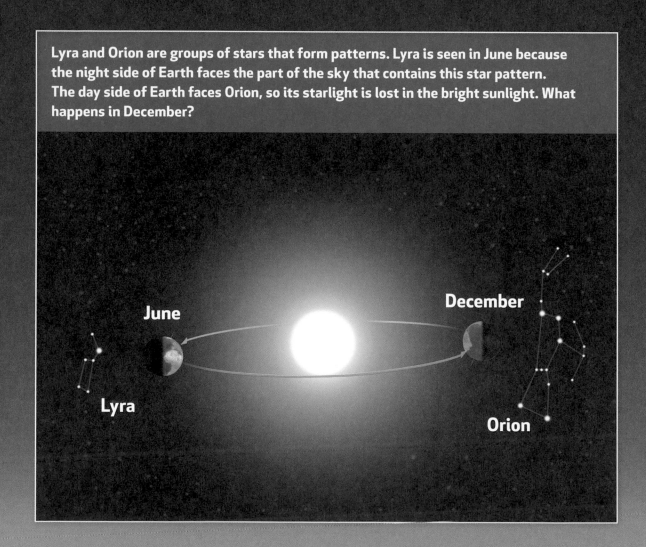

Tools such as telescopes and binoculars help you see dim and far away stars and other objects. Many telescopes use lenses to gather starlight and magnify stars, or make them appear larger. If you look at a patch of starless sky through a telescope or binoculars, stars will become visible. That's because the lenses in the tool are much larger than the lenses in your eyes. Other telescopes use mirrors that gather light in a similar way to see far away stars.

Scientists at the Lowell Observatory in Flagstaff, Arizona, have the moon in their sights.

Even small telescopes let you see stars you can't see with your eyes alone.

Observing the Moon The moon is the biggest and brightest object in the night sky. It is also the second brightest object in the daytime sky. Yet you can see more details about the moon if you observe it through a telescope or binoculars.

You can learn even more about a place by actually going there. Scientists have sent about 70 spacecraft to orbit or land on the moon. Those in orbit take photographs. Spacecraft that land can scoop up shovelfuls of soil and conduct experiments. Some spacecraft are crashed on purpose! Then other tools can measure the materials that are kicked up from the surface.

A small telescope or pair of binoculars lets you see details of the moon like these.

What have people learned from observing the moon? This rocky world has no air, almost no water, and no life. Its rocky, dusty surface is marked with craters. These bowl-shaped dents are created when rocks traveling through space smash into the moon. Large dark spots also cover parts of the moon's surface. People used to think they were seas. Now we know they are plains, or lowlands, of lava rock. The light areas on the moon are hills and mountains, or highlands.

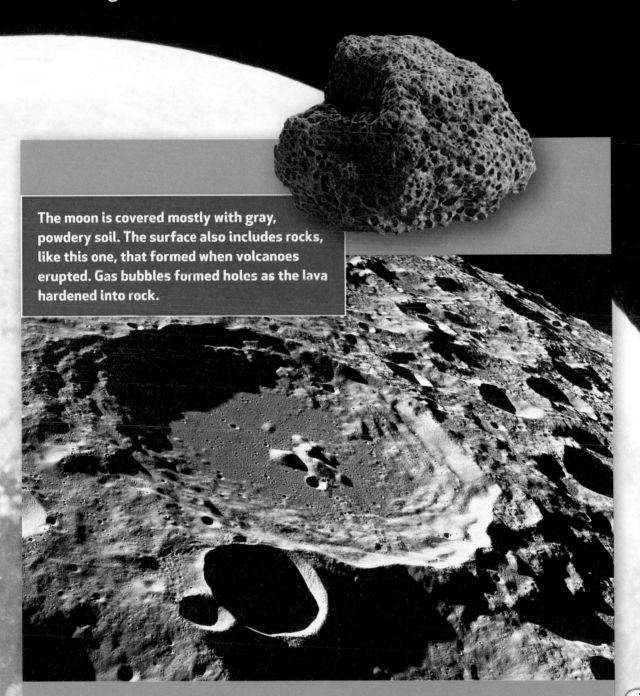

The moon is covered mostly with gray, powdery soil. The surface also includes rocks, like this one, that formed when volcanoes erupted. Gas bubbles formed holes as the lava hardened into rock.

The moon is the only object in space that people have actually visited. Twelve astronauts have walked on the moon. The moon has weak gravity, so walking on its surface is easy.

Space suits protected the astronauts from the moon's extreme temperatures. Where the sun shines, temperatures can get twice as hot as the hottest desert on Earth. At night, it gets almost three times colder than the coldest places in Antarctica.

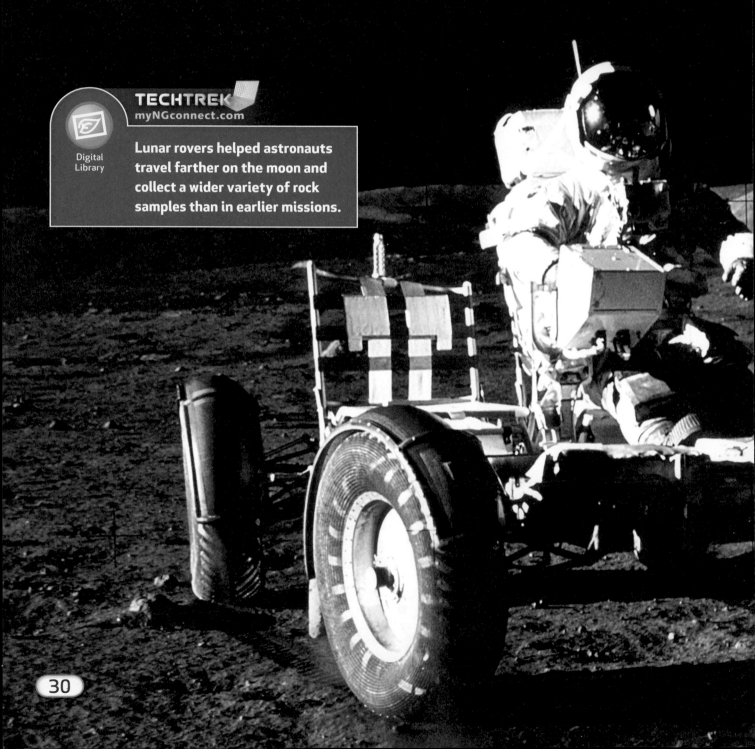

TECHTREK
myNGconnect.com

Digital Library

Lunar rovers helped astronauts travel farther on the moon and collect a wider variety of rock samples than in earlier missions.

Some astronauts drove lunar rovers to explore the moon. They observed the land features and collected rocks. The rocks show what the moon is made of. They also show how old the moon is. Scientists think the moon formed about 4.5 billion years ago because that is the age of the oldest rocks collected.

An astronaut examines a space probe that landed on the moon two years earlier.

Before You Move On

1. How do telescopes help people observe stars and the moon?
2. What causes craters?
3. **Evaluate** What evidence do scientists use to determine the age of the moon?

The Moon Moves Around Earth

Just as Earth revolves around the sun, the moon revolves around Earth. The moon completes its revolution around Earth about every four weeks. Gravity between Earth and the moon holds the moon in orbit. The moon also travels with Earth in its orbit around the sun.

Just as Earth rotates on an axis, the moon does too. So this space traveler has three motions: it rotates, it revolves around Earth, and, with Earth, it revolves around the sun.

 The moon is always on the move, but it appears to rise in the sky because of Earth's motion.

The moon rotates once in about four weeks, the same amount of time it takes to revolve around Earth. For that reason, the same side of the moon always faces Earth.

The moon may be the brightest object in the night sky, but it does not give off its own light. It reflects light from the sun. The "moonlight" you see is sunlight that has bounced off the moon.

The sun shines on the half of the moon that is facing it just as the sun shines on the half of Earth that faces it. As the moon revolves around Earth, you see different amounts of its lighted half.

You can show how the moon rotates and revolves in the same amount of time. A penny represents the moon and a quarter represents Earth.

Lay both coins on a surface. Move the penny around the quarter so that Lincoln's nose always points toward the quarter. In order to do that, you have to rotate the penny on its axis as the penny revolves around the quarter.

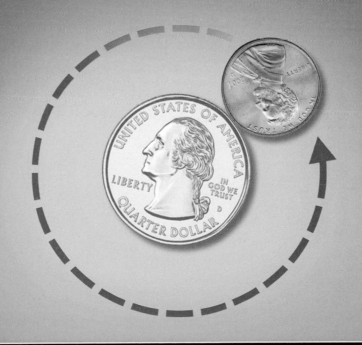

Phases of the Moon

Observe the moon over several nights. It looks like it changes shape in the sky. Sometimes it's a full circle and other times it's a half circle. But the moon doesn't really change shape. It's always shaped like a sphere, or ball. What changes is how much of its lighted half you can see from Earth. How much of the lighted half you see depends on where the moon is in its revolution around Earth.

TECHTREK
myNGconnect.com

Digital Library

For this month, on which day do you start seeing less and less of the moon?

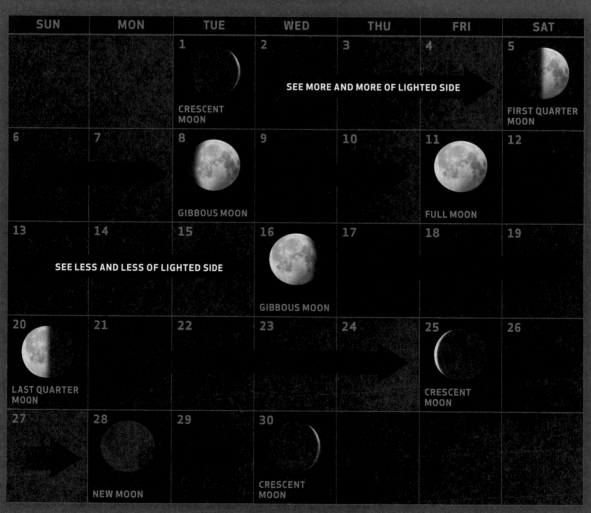

SUN	MON	TUE	WED	THU	FRI	SAT
		1 CRESCENT MOON	2	3	4	5 FIRST QUARTER MOON
6	7	8 GIBBOUS MOON	9	10	11 FULL MOON	12
13	14	15	16 GIBBOUS MOON	17	18	19
20 LAST QUARTER MOON	21	22	23	24	25 CRESCENT MOON	26
27	28 NEW MOON	29	30 CRESCENT MOON			

SEE MORE AND MORE OF LIGHTED SIDE

SEE LESS AND LESS OF LIGHTED SIDE

On some nights you can see the entire lighted side of the moon. This is a full moon. On other nights you can see most of the lighted side, half of it, a sliver of it, or none of it. As the moon revolves around Earth, you see changing amounts of the lighted side. So the moon seems to change shape. Each different shape is a **phase** of the moon.

It takes the moon about four weeks to go through all of its phases. The pattern of phases repeats itself. So you can predict what the moon will look like tomorrow, next week, and even a year from now.

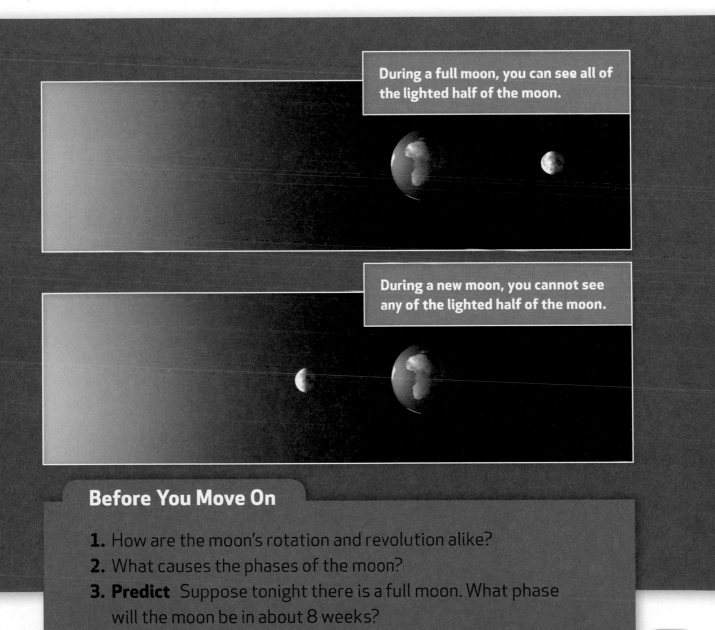

During a full moon, you can see all of the lighted half of the moon.

During a new moon, you cannot see any of the lighted half of the moon.

Before You Move On

1. How are the moon's rotation and revolution alike?
2. What causes the phases of the moon?
3. **Predict** Suppose tonight there is a full moon. What phase will the moon be in about 8 weeks?

TELLING TIME
YESTERDAY AND TODAY

What month is it? You would likely check a calendar to find out. What time is it? You would probably look at a clock. Long ago, people noticed the patterns in the motions of Earth, the moon, and the sun. They organized time according to these patterns and invented calendars and clocks based on them.

The calendar people use today divides the 365-day year into 12 months. Each month has 28 to 31 days. That is close to the length of the moon's phases, but not exactly. Days are grouped into weeks. The 7-day week is not based on any pattern of motion in the sky.

Calendars People in ancient cultures noticed the pattern of sunrise and sunset each day. They also saw that the position of the sun changed in the sky through the seasons. By observing and counting, they discovered it took 365 days for the sun to return to a certain position. Making a calendar helped people track the seasons and the years.

The **Aztec** of early Mexico created a calendar based on 365 days in a year. Each year had 18 months; each month had 20 days. That left 5 extra days at the end of the year.

Long ago at Stonehenge in England, people carefully placed these rocks. Scientists don't know exactly why Stonehenge was built. But they do know that the stones are arranged in such a way that they mark the beginning of the seasons. For example, the sun rises directly over a certain rock here on the first day of summer.

Clocks People further organized time by dividing a day into 24 hours. Each hour has 60 minutes. Each minute has 60 seconds. Inventing clocks helped people keep track of the time during each day.

Modern clocks and watches show time to the minute. Stopwatches show seconds and even tenths or hundredths of seconds.

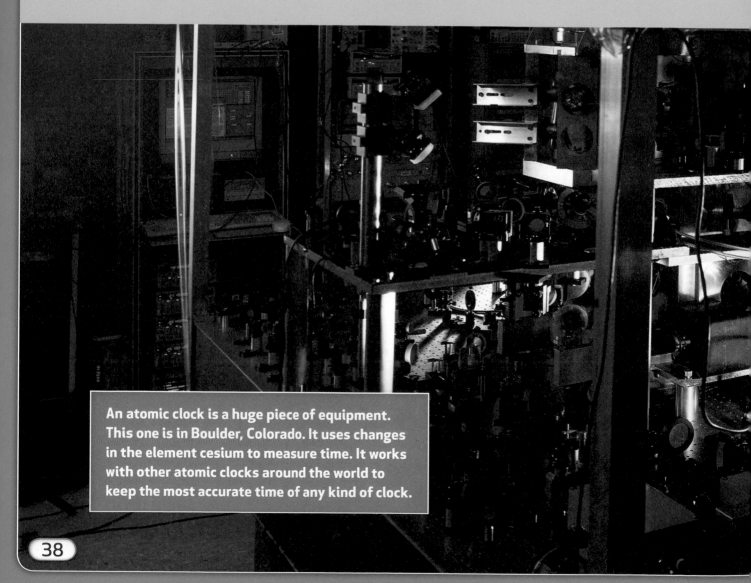

An atomic clock is a huge piece of equipment. This one is in Boulder, Colorado. It uses changes in the element cesium to measure time. It works with other atomic clocks around the world to keep the most accurate time of any kind of clock.

Leap Year

Earth actually takes a little more than 365 days to revolve around the sun. It takes 365 ¼ days. Calendars can't just ignore this fact. Every four years, that extra quarter day adds up to one full day. After many years, the calendar would no longer match up to the seasons. So every four years, a day is added to the calendar year. This "leap year" has 366 days instead of 365.

February usually has 28 days. In a leap year it has 29 days.

Conclusion

Earth and its moon are part of the solar system. Earth's daily rotation causes day and night and makes the sun, moon, and some stars appear to move across the sky. Earth's yearly revolution causes you to see different stars at different times of the year. The moon also rotates and revolves, but around Earth. Its rotation and revolution take about the same length of time. The moon's revolution around Earth results in phases of the moon.

Big Idea Earth and its moon both rotate, and Earth revolves around the sun while the moon revolves around Earth.

SUN

EARTH

MOON

Cycles

Vocabulary Review

Match the following terms with the correct definition.

A. phase
B. latitude
C. gravity
D. rotate
E. revolve
F. axis

1. To spin around
2. An imaginary line around which Earth spins
3. A lighted part of the moon as it appears from Earth
4. A force that pulls objects toward each other
5. How far north or south of the Equator a place is
6. To move around another object

Big Idea Review

1. **Identify** What kind of tools help people see objects that are very far away?

2. **Restate** What kinds of movements does the moon make about once each month?

3. **Compare and Contrast** How are Earth and the sun and moon alike and different?

4. **Cause and Effect** What causes the sun, moon, and stars to seem as though they are moving across the sky?

5. **Apply** If you said the Earth's revolution around the sun started in March, about how far around the sun is Earth by December?

6. **Draw Conclusions** Suppose you are standing in the shadow of a tree. The shadow is long and stretches to the east. About what time of day is it? Explain how you can tell.

my **SCIENCE** notebook

Write About Earth's Movement

Apply Imagine you are in Antarctica observing this moon phase. Explain where the sun, Earth, and moon are in relation to one another. Tell what is happening with the sunlight that enables you to see this moon.

EARTH SCIENCE EXPERT: CULTURAL ASTRONOMER

What do different cultures think about the sky?

Everyone can see the sky. But the night sky looks different to people depending on where they live. People from different cultures think about the sky differently too. Jarita Holbrook, a cultural astronomer, finds that fascinating. She travels around the world talking to people. She asks them what they know about the sky. She also looks at how they use their knowledge. "Everybody has something interesting to say about the sky," she says.

Holbrook traveled to Africa when a solar eclipse occurred. That's when the moon passes directly between Earth and the sun, blocking out the sun for a couple minutes. She wanted to see how the people responded to this event. Another time she traveled to the Fiji islands in the Pacific Ocean. There, she learned how people use the stars to navigate, or find their way traveling on the water. Holbrook has studied African folktales and myths, too. Why? Because they show the people's knowledge of the moon and sun.

Jarita Holbrook is interested in what people from different cultures think about what they see in the sky.

TECHTREK
myNGconnect.com

Student
eEdition

Digital
Library

TECHTREK
myNGconnect.com

Digital
Library

During a full solar eclipse, the moon blocks out the sun for several minutes.

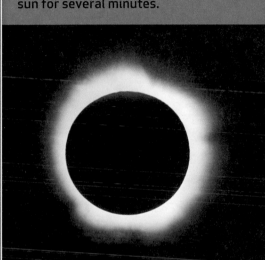

Cultural astronomers such as Holbrook study what people today think about the sky. Other cultural astronomers study how long-ago peoples viewed the sky.

Do you like learning about people and the night sky? If you said yes, Holbrook thinks you might like being a cultural astronomer. Holbrook studied science and math in school. Later she became interested in cultural astronomy. What if you wanted to become a cultural astronomer? Holbrook recommends studying subjects about cultures, such as anthropology or archaeology. And, of course, the night sky!

Observers need special glasses to view a solar eclipse.

BECOME AN EXPERT

The Earth-Moon-Sun System:
How Knowledge Grows

People used to think Earth was at the center of the universe. It is easy to understand why. When you watch the sky, the sun, the moon, and many stars appear to rise toward the east and set toward the west. They seem to circle around Earth, and Earth seems to stand still.

You know that Earth **revolves** around the sun, not the other way around. The moon revolves around Earth and travels with Earth around the sun. Over thousands of years, people from many cultures and countries contributed to this understanding of movements in space.

revolve

To **revolve** is to move around another object.

TECHTREK
myNGconnect.com

Student
eEdition

Digital
Library

Ancient Discoveries

People have long been interested in space and Earth's place in it. You may have learned that during the days of Christopher Columbus some people thought Earth was flat. Actually scientists had discovered Earth's true shape thousands of years before. Ancient Greeks showed that Earth is round and used math to calculate its size. Arab astronomers built observatories to watch objects in the sky. They also improved tools for observation and wrote books on astronomy, or the study of space. Al-Battani was the most famous Arab astronomer. He figured how long it takes Earth to orbit the sun.

Arab astronomers mapped the stars on globes like this one.

BECOME AN EXPERT

In ancient India, the astronomer Aryabhata figured out the shape of the planets' orbits. Aryabhata suggested that Earth **rotates** on an **axis**. He also realized that the moon reflects light and does not shine on its own.

More than 3,000 years ago, the Chinese created a calendar based on both the sun and the moon's motion. Serving their emperor, Chinese astronomers measured the tilt of Earth, mapped the stars, and invented tools for observing the sky.

Aryabhata wrote that objects only seem to move through the sky because Earth rotates. Later scientists changed that idea in his book because they thought it was a mistake.

The Chinese built on their ancient knowledge and constructed this sundial sometime during the 1400s or 1500s.

rotate
To **rotate** is to spin around.

axis
An **axis** is an imaginary line around which Earth spins.

The ancient Maya lived in Mexico and Central America. The Maya observed the sun's movements closely and figured out the length of a year. Their pyramids and other buildings used sunlight and shadow to mark events such as the beginning of spring.

TECHTREK
myNGconnect.com

Digital Library

The Mayan calendar used symbols to represent periods of time throughout the year.

El Caracol is an ancient Mayan observatory at Chichen Itza.

Advances in Europe

Copernicus and Galileo Some of the discoveries made long ago were not well known in Europe in the 1400s. At that time, most people believed Earth was the center of the universe and all objects in space revolved around Earth. That didn't make sense to Nicolaus Copernicus, a Polish scientist who lived in the 1500s. He suggested that Earth and all the other planets revolve around the sun. Copernicus said Earth rotates and that its rotation made the sun appear to move across the sky.

Later, scientists made discoveries that supported Copernicus's ideas. The Italian scientist Galileo Galilei used the newly invented telescope to discover four moons orbiting Jupiter.

This proved that all objects in space did not orbit Earth. Galileo also observed that Venus goes through **phases** just like the moon. These phases would only be possible if Venus orbited the sun.

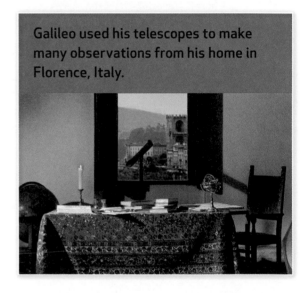

Galileo used his telescopes to make many observations from his home in Florence, Italy.

Copernicus's hypothesis that Earth revolved around the sun upset many Europeans at the time.

phase

A **phase** is a lighted part of the moon as it appears from Earth.

Kepler and Newton The German astronomer Johannes Kepler refined ideas about the sun-centered system. He showed that the orbits were ellipses, not perfect circles like other people thought. But he didn't know why the planets stayed in orbit at all. Scientists spoke of a "planet-moving force," but no one knew what it was or how it worked.

That changed in 1666. In that year, Sir Isaac Newton described this force as **gravity**. He explained that every object in the universe attracts every other object. The strength of this force of gravity is greater for objects that have more mass and are closer together. He showed that without gravity, planets would shoot off into space in straight lines. But gravity keeps them in orbit around a more massive body—the sun. Newton used ideas of earlier scientists to finally explain the Earth-moon-sun system.

English scientist Isaac Newton supported the ideas of Copernicus and Galileo with his theory of gravity.

gravity

Gravity is a force that pulls objects toward each other.

Modern Technology

Technology has come a long way since Newton's day. Over the years, scientists and engineers have built better and bigger telescopes that can see deeper and deeper into space. Some of these telescopes are in orbit around Earth, above the air that interferes with viewing. People have sent spacecraft not only to the moon but to other planets and beyond. Today close-up pictures show planets, moons, and other objects that people couldn't even see in the sky a hundred years ago.

This observatory near La Serena, Chile, includes several domed buildings. Each holds a huge telescope. The roof of each building opens to allow viewing of the sky.

Astronomers are learning more about the Earth-moon-sun system. But they are also peering into other systems far away. They are discovering planets around other stars. Who knows what they might find.

This astronomer attaches a tool to a telescope that lets her observe evidence of over 100 far away star systems at one time!

CHAPTER
1

SHARE AND COMPARE

Turn and Talk How have ideas about the sun, Earth, and the moon changed over time? Form a complete answer to this question together with a partner.

Read Select two pages in this section. Practice reading the pages. Then read them aloud to a partner. Talk about why the pages are interesting.

Write Write a conclusion that tells the important ideas of what you have learned about how knowledge of the Earth-sun-moon system has grown. State what you think is the Big Idea of this section. Share what you wrote with a classmate. Compare your conclusions. Did your classmate recall that many ancient cultures had ideas that matched what others proved later?

Draw Draw a picture that shows the idea of an ancient culture or another person in history about the relationship of the sun, Earth, and the moon. Combine your drawings with those of your classmates to create time lines that show how knowledge grows.

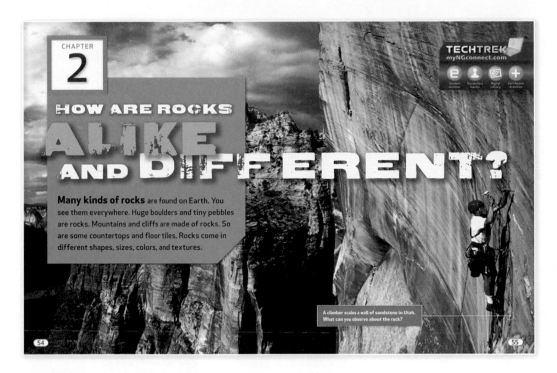

CHAPTER
2

HOW ARE ROCKS
ALIKE
AND DIFFERENT?

Many kinds of rocks are found on Earth. You see them everywhere. Huge boulders and tiny pebbles are rocks. Mountains and cliffs are made of rocks. So are some countertops and floor tiles. Rocks come in different shapes, sizes, colors, and textures.

TECHTREK
myNGconnect.com

A climber scales a wall of sandstone in Utah. What can you observe about the rock?

54

55

After reading Chapter 2, you will be able to:

- Recognize that most rocks are made of minerals. **ROCKS AND MINERALS**

- Identify and compare properties of rocks. **ROCKS AND MINERALS**

- Identify and compare properties of minerals. **PROPERTIES OF MINERALS**

- Identify and compare the three categories of rocks: igneous, sedimentary, and metamorphic. **TYPES OF ROCKS**

- Science in a Snap! Identify and compare the three categories of rocks: igneous, sedimentary, and metamorphic. **TYPES OF ROCKS**

CHAPTER 2

HOW ARE ROCKS ALIKE AND DIFF

Many kinds of rocks are found on Earth. You see them everywhere. Huge boulders and tiny pebbles are rocks. Mountains and cliffs are made of rocks. So are some countertops and floor tiles. Rocks come in different shapes, sizes, colors, and textures.

TECHTREK
myNGconnect.com

Student
eEdition

Vocabulary
Games

Digital
Library

Enrichment
Activities

ERENT?

A climber scales a wall of sandstone in Utah.
What can you observe about the rock?

SCIENCE VOCABULARY

mineral (MIN-u-ruhl)

A **mineral** is a solid, nonliving material that forms in nature. (p. 59)

This purple mineral is called amethyst.

property (PROP-ur-tē)

A **property** is something about an object that you can observe with your senses. (p. 60)

A property of this rock is that it has many holes.

grain (GRĀN)

Grains are small mineral or rock pieces. (p. 61)

The grains in this brown sandstone are large enough to see.

my

Science Vocabulary

grain
(GRĀN)

igneous rock
(IG-nē-us ROK)

property
(PROP-ur-tē)

metamorphic rock
(met-a-MOR-fik ROK)

mineral
(MIN-u-ruhl)

sedimentary rock
(sed-i-MEN-tah-rē ROK)

TECHTREK
myNGconnect.com

Vocabulary
Games

igneous rock
(IG-nē-us ROK)

Igneous rock is rock that forms when melted rock cools and hardens. (p. 70)

As this lava cools and hardens, it forms a type of igneous rock.

sedimentary rock
(sed-MEN-tah-rē ROK)

Sedimentary rock is rock that forms when small pieces of rock and other materials settle and gel squeezed or cemented together. (p. 72)

This sedimentary rock formed when desert sand settled and hardened into sandstone.

metamorphic rock
(met-a-MOR-fik ROK)

Metamorphic rock is rock that has been changed by heat or pressure. (p. 76)

SHALE → SLATE

Pressure and heat changed this shale to slate, a metamorphic rock.

Rocks and Minerals

What do you picture when you hear the word *rock*? Just something dull and gray? Just something to walk on, climb on, or kick out of the way? You may be surprised at the variety of rocks and how interesting they can be.

This rock in Antelope Canyon, Arizona, is made up of minerals.

Coal is an unusual rock. It isn't made of minerals at all. Coal formed from plants that lived millions of years ago.

The purple mineral amethyst is often used in jewelry.

If you observe rocks carefully, you will notice that they have different **properties**. A property is something about an object that you can observe with your senses.

Scientists classify rocks using some of the properties in this chart. You can use these properties to observe and compare the rocks you see around you.

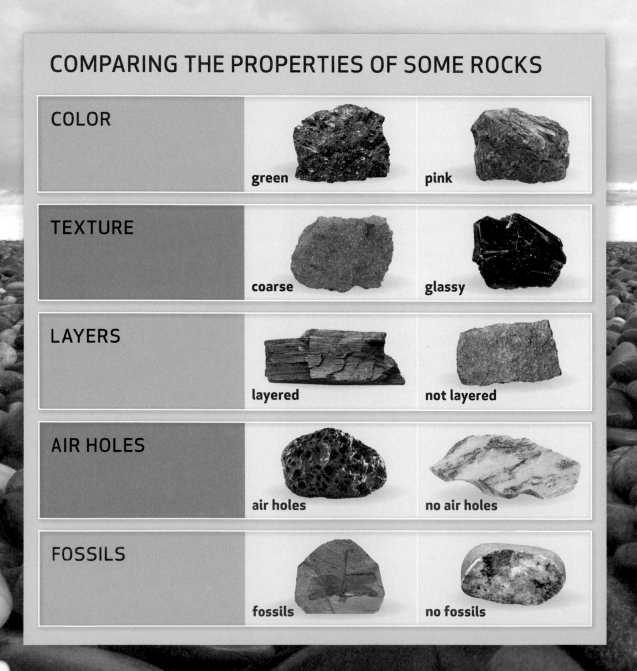

COMPARING THE PROPERTIES OF SOME ROCKS

COLOR	green	pink
TEXTURE	coarse	glassy
LAYERS	layered	not layered
AIR HOLES	air holes	no air holes
FOSSILS	fossils	no fossils

If you look at a rock through a hand lens, you might see its **grains**. Grains are pieces of minerals that make up the rock. Rocks with larger grains are coarser than rocks with smaller grains. Rocks with no grains or grains too small to be seen through a hand lens are glassy.

The shape and size of rocks are properties too. You can use these properties to describe rocks. Changes in the environment can change the shape and size of rocks. For example, water in a stream can wear away sharp rocks into smooth pebbles. The kinds of minerals that make up rocks also can affect the rocks' shape and size.

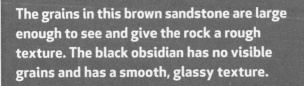

The grains in this brown sandstone are large enough to see and give the rock a rough texture. The black obsidian has no visible grains and has a smooth, glassy texture.

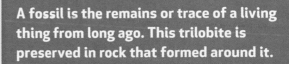

A fossil is the remains or trace of a living thing from long ago. This trilobite is preserved in rock that formed around it.

Before You Move On

1. How are all rocks alike?
2. Give three properties of rocks.
3. **Infer** If the rock looks shiny like a piece of glass, what can you infer about its grain size?

Properties of Minerals

The properties of rocks come from the minerals that make them up. Copper, quartz, and lead are just a few of the nearly 4,000 known minerals. How do scientists tell them apart? They observe their different properties.

Color The first property you might notice about a mineral is its color. However, some minerals come in many different colors. Also, many different minerals have the same color. So color alone cannot be used to identify most minerals.

The mineral quartz comes in several colors.

Streak Streak is better than color for telling minerals apart. Streak is the color of powder a mineral leaves when rubbed on a rough tile. A mineral may come in many colors, but the color of its streak is always the same. Some minerals do not leave any streak because they are harder than the tile.

Pyrite can look so much like gold that it is called "fool's gold." It cannot fool anyone who looks at its streak. Gold has a gold streak and pyrite has a black streak.

Gold

Pyrite

Hardness Scientists measure hardness by how easily the mineral can be scratched. You may think all minerals are hard. But you can scratch some with your fingernail!

Scientists use the Mohs scale to test and measure hardness. Look at the minerals in the chart below. Talc is the softest mineral. It can be scratched by all the other minerals on the chart. At the other end of the scale is diamond. Diamond is the hardest mineral. It can scratch all the other minerals on the chart.

MOHS SCALE OF HARDNESS

HARDNESS				
1	**2**	**3**	**4**	**5**
QUICK TEST Scratched easily by a fingernail	Scratched by a fingernail	Scratched by a penny	Scratched by an iron nail	Scratched by glass
MINERAL TALC	GYPSUM	CALCITE	FLUORITE	APATITE

You don't need a set of the ten minerals in the Mohs scale to test for hardness. Look at the chart again. Find the row titled *Quick Test*. You will find a list of objects you can use to test for hardness. For example, if a mineral can be scratched by a piece of glass but not by an iron nail, the mineral's hardness is between 4 and 5. If you can scratch the mineral with your fingernail, the mineral has a hardness between 1 and 2.

6	7	8	9	10
Scratched by a steel file	Can scratch steel and glass	Can scratch quartz	Can scratch topaz	Can scratch all other substances
ORTHOCLASE	QUARTZ	TOPAZ	CORUNDUM	DIAMOND

Other Properties
Minerals can have other properties too. These include cleavage, magnetism, luster, and reacting with acid.

Cleavage   Some minerals break, or cleave, along smooth flat surfaces. This property is called cleavage. Pyrite, for example, has cleavage because it can break into cubes. Mica has cleavage because it breaks into smooth, flat sheets.

Magnetism   Some minerals act like magnets. They attract objects that contain iron. One of the most magnetic minerals is magnetite. An old story tells how a shepherd in the area of Greece called Magnesia discovered magnetite. The nails in his sandals stuck to the rock beneath his feet.

Mica breaks into thin, flat sheets. It has good cleavage.

Magnetite is magnetic. It attracts paperclips, staples, and nails.

Luster Luster is how a mineral's surface reflects light. Luster can be metallic or nonmetallic. A mineral with a metallic luster reflects light like a shiny piece of metal. Hematite has a metallic luster because it contains metals. Minerals with a nonmetallic luster may look glassy, pearly, greasy, or dull. Compare the lusters of the minerals shown here.

Reacts with Acid Some minerals bubble or fizz when they come in contact with a weak acid, such as vinegar. Calcite has this property.

Vinegar was placed on this rock containing calcite. The calcite fizzes as it releases gas bubbles.

DIFFERENT KINDS OF LUSTER

Hematite can have a metallic luster.

This beryl has a glassy luster.

Talc's luster is greasy, or oily.

Malachite's luster is dull.

Minerals in Rocks

Minerals are the building blocks of rocks. Minerals give rocks their unique characteristics. Many minerals are very small and hard to see. Scientists can use tools such as a hand lens or a microscope to look closely at rocks to see the minerals they contain.

Some rocks are made of only one mineral. The rocky chalk cliffs in the photo are made only of the mineral calcite. Sidewalk chalk used to be made of this mineral. Have you ever seen rock salt spread on an icy street? Rock salt is made of the mineral halite.

These chalk cliffs are made of the mineral calcite.

You need a microscope to see the crystal structure of the calcite.

Most rocks are made of more than one mineral. The rock granite is made mostly of quartz and feldspar. It also contains small amounts of mica and other minerals. These minerals used to be part of a slushy mixture deep underground. As the mixture hardened, the minerals locked together. They formed hard, solid granite.

A hand lens helps you see the minerals in granite. You can tell the minerals apart by their different shapes and colors.

Mica

Feldspar

Quartz

Before You Move On

1. List three properties of minerals other than color.
2. Why is color alone not a good way to identify minerals?
3. **Apply** Imagine you have a mineral that is scratched by glass but not by a penny. Which one of these minerals could it be—gypsum, calcite, or apatite?

Types of Rocks

Scientists classify rocks into three basic types. Each type forms in a different way.

Igneous Rock Igneous rock forms when melted rock cools and hardens. Deep below Earth's surface, temperatures are high enough to melt rock. This gooey molten rock is called magma. It is less dense, or lighter, than the material around it, so it rises toward the surface. Sometimes magma escapes through openings in Earth's surface called volcanoes. At the surface, magma is called lava. The lava cools rapidly, which makes the grains in the rock small and hard to see.

TECHTREK
myNGconnect.com

Enrichment Activities

Rivers of red-hot lava pour from this Hawaiian volcano.

Lava usually hardens quickly, forming a type of igneous rock.

Most igneous rocks form underground. The magma rises slowly and stops below the surface. Since the magma does not come in contact with air or water, it cools and hardens very slowly. During slow cooling, the grains grow large enough to see. The igneous rock granite forms from this slow cooling of magma. It has large grains.

Over millions of years, forces in Earth lift the granite above the surface. Mountains form. The Andes Mountains in South America are made of granite that formed underground.

These mountains in Chile are made of igneous rock that formed slowly, below Earth's surface. The Hawaiian Islands are made of igneous rock that formed quickly, above Earth's surface.

Sedimentary Rock One way **sedimentary rocks** form is when small pieces of rock and other materials settle on top of each other and become squeezed together. These pieces, called sediments, build layer upon layer. New layers press down on old layers. Some minerals act like cement and hold the sediments together. That's how the sedimentary rock called sandstone forms. The sand might come from deserts, beaches, and the bottom of shallow seas. Over millions of years, the layers become sandstone.

These sand dunes form layers of sediment.

The lower layers are squeezed together by the weight of the upper layers. Minerals between the sand hold the grains together.

What makes layers of sedimentary rock look different from each other? It's often the color. Different amounts of the mineral iron in the rock may make it look yellow, orange, red, or black. Sometimes different layers are different kinds of sedimentary rock. Chalk, rock salt, and other kinds of limestone are made of sediments that are different than the sediments in sandstone.

Overtime, wind, water, and ice uncover the sandstone. You can see where the separate layers of sediment hardened.

Some sandstone in Zion National Park was made from layers of sand that collected at the bottom of a shallow sea that once covered the area.

Fossils in Sedimentary Rocks Did you know you can observe history in rocks? Sedimentary rocks often contain fossils. They give scientists clues to the past. For example, a fossil such as a dinosaur tooth can help scientists tell when the rock formed. A fossil of a fish on land can show that the land was once covered by water. In a desert, the fossils of rain forest plants would show that the climate used to be much wetter.

TECHTREK
myNGconnect.com

Digital Library

The fossil of a huge scallop shell is a clue to scientists that the area was covered by water millions of years ago.

Put a layer of sand in a cup. Put a "fossil" paper clip in the sand.

Add layers of coffee grounds and salt onto the layer of sand, adding other "fossil" rubber bands and counters as you do so.

Which layer is the oldest?

Where would you expect to find the oldest fossil?

A scientist uncovers fossils from rock along the coast of Rhode Island.

Metamorphic Rock After rock forms, it can change into any other type of rock. Metamorphic rock forms when existing rock is changed by heat or pressure. The mineral crystals in the rock can change shape or arrange themselves in different patterns. This changes the properties of the rock. For example, pressure can change the sedimentary rock limestone into the metamorphic rock marble, a much harder rock.

SEAFLOOR SEDIMENT

high pressure

LIMESTONE high heat

MARBLE

Marble from this quarry in Carrara, Italy, is prized by sculptors for its even white color.

Every kind of igneous and sedimentary rock can change by heat or pressure into a certain kind of metamorphic rock. For example, sandstone is a kind of sedimentary rock. Heat and pressure can change sandstone to quartzite. Quartzite is much harder and smoother than sandstone. Sometimes pressure causes mineral grains to arrange themselves into layers, like lots of sheets of paper stacked one on top of the other. The metamorphic rock, slate is made up of layers. It can be easily split apart.

MUD → SHALE (high pressure, high heat) → SLATE

SAND → SANDSTONE (high pressure, high heat) → QUARTZITE

Before You Move On

1. Describe how sediment in a shallow sea can form marble.
2. If sandstone melts and then hardens, what does it become?
3. **Infer** Imagine you have one type of igneous rock with grains too small to see, and another type of igneous rock with grains you can see easily. What can you infer about how the rocks formed?

DIAMONDS
AND THEIR USES

Minerals that are especially rare and beautiful are called gems. Among the most prized of all gems is diamond. This mineral forms in magma under extreme heat and pressure. Most diamond crystals are found in the narrow openings of ancient volcanoes. Natural diamond crystals are not much to look at. If you found one, you might just toss it away. But when the crystal is cut and polished the right way, its brilliant luster really shines.

Most diamonds are either too small or not clear enough to make beautiful jewelry. But even these diamonds are useful. That's because diamond is the hardest material known. So diamonds are used in tools for cutting, drilling, and grinding. You can even buy a diamond nail file. What can be used to polish something as hard as a diamond? Would you believe diamond dust?

Diamonds have long been used in crowns, rings, and all kinds of jewelry.

What tool can cut through a street? A saw coated with bits of diamond can.

Raw diamonds look more like dirty, greasy stones than brilliant jewels.

Jewelers figure out the best way to cut diamonds to bring out their brilliance.

Conclusion

Many rocks and minerals are found on Earth. Both rocks and minerals have different properties. Properties help tell them apart. Most rocks are made of minerals. Scientist sort rocks into three types according to how they formed. These types are igneous, sedimentary and metamorphic.

Big Idea Rocks can be sorted depending on the minerals in them, their properties, and how they formed.

TYPES OF ROCK

igneous — GRANITE

sedimentary — SANDSTONE

metamorphic — MIGMATITE

Vocabulary Review

Match the following terms with the correct definition.

A. **igneous rock**

B. **metamorphic rock**

C. **sedimentary rock**

D. **property**

E. **mineral**

F. **grain**

1. Something about an object that you can observe with your senses

2. A solid, nonliving material that forms in nature.

3. A particle of a mineral or rock.

4. Rock that forms when small pieces of rock and other materials settle and get squeezed or cemented together.

5. Rock that has been changed by heat or pressure

6. Rock that forms when melted rock cools and hardens

Big Idea Review

1. **Describe** Name a type of rock and list its properties.

2. **Define** What is streak?

3. **Explain** How do scientist identify very small minerals?

4. **Compare and Contrast** How are igneous rock and sedimentary rock alike and different?

5. **Infer** The igneous rock granite forms beneath Earth's surface. The igneous rock obsidian forms on the surface. Which has the smoother texture? Why?

6. **Generalize** How might metamorphic rock change into igneous rock?

Write About How Rocks Form and Change

Interpreting Diagrams Look at these photos. Explain how rock is forming and changing.

SAND

SANDSTONE

QUARTZITE

EARTH SCIENCE EXPERT: SCULPTOR

How Can an Artist Use Science? Ask a Sculptor.

Art may be one of your hobbies. How can science be helpful to artists? Sculptor Edward Fleming uses his knowledge of earth science when working with stone.

As a sculptor, what do you do?

I am an artist. I carve my art in stone.

Is your work strongly connected to earth science?

Definitely. I think about how stone is formed. What I find inside tells the history of Earth. I mostly use limestone, which is sedimentary, and marble, which is metamorphic. Millions of years ago, sea creatures died and left their remains on the sea bed. Over time, these remains formed limestone. Pressure and heat changed the limestone into marble. In the Rocky Mountains, the marble was lifted. It is now more than a mile above sea level. I live and work near this marble.

TECHTREK
myNGconnect.com

Digital Library

This sculpture of a human face was carved, by Fleming, from limestone.

TECHTREK
myNGconnect.com

Student
eEdition

Digital
Library

What properties of marble affect your sculptures?

Color, light, crystal structure, and veins of material affect my sculptures.
I also think about how the stone breaks.

How could studying science help with sculpture?

It is important to understand your material. I know a lot about stone.
Science also helps you understand your subject. I sculpt the human body.
That is why I have studied anatomy.

How did you study to become a sculptor?

I studied sculpture and other arts. Studying architecture also helped me.

What advice would you give students who want to be sculptors?

Put your hands on the material. It can be clay, papier-mâché, or stone.
Start working with it. Learn as you work. Learn from your mistakes.

Fleming sculpted this monument for
a community in New Mexico.

83

BECOME AN EXPERT

The Grand Canyon:
History Written in Rock

The Grand Canyon is an awesome wonder. No pictures can show you how truly huge it is. The canyon is 1,524 meters (5,000 feet) deep. That is about as high as a stack of 450 school buses! Parts of it are 29 kilometers (18 miles) wide. It would take a car moving at 55 miles per hour nearly 20 minutes to drive across it.

Running water shapes the Grand Canyon. The Colorado River began carving the canyon about six million years ago. It cut through many layers of **sedimentary rock**, forming the spectacular walls.

The Colorado River has carved its way through almost a mile of rock layers, forming the Grand Canyon.

sedimentary rock

Sedimentary rock is rock that forms when small pieces of rock and other materials settle and get squeezed or cemented together.

One of the best ways to see the canyon is by mule. The sure-footed animals carry visitors from the rim to the bottom. Let's go along on one of these rides. As you'll see, it's a trip back in time.

At a signal from the guides, the mules begin their journey. They plod past the first of many rock layers. The deeper we go, the older the layers of rock will be. The guides tell us to look at the layers carefully. The layers will change color because they are made of different kinds of sedimentary rock. Color is one **property** of rock.

A mule trip offers dramatic views of the canyon walls.

property

A **property** is something about an object that you can observe with your senses.

We sway with the movement of the mules. The guide points out one of the top layers. It is limestone. This sedimentary rock formed about 240 million years ago. Limestone is made mostly of the **mineral** calcite. The mineral comes from the shells of sea animals. This is evidence that an ocean once covered the area. Looking around at the desert landscape, it's hard to imagine that this all used to be under water, but it's true.

The guide points out a fossil in the canyon walls. It is a fossil of an animal that lived here millions of years ago. Scientists can use fossils to help them estimate the age of a rock layer.

This fossil shell was found in the canyon's limestone. It is evidence that sea animals lived in the area over 250 million years ago.

This fossil shell shows that clam-like animals once lived here.

mineral

A **mineral** is a solid, nonliving material that forms in nature.

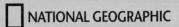

The mules follow a trail that zigzags down the canyon. Now the cliffs are made of sandstone, which formed from sand. Desert sand covered this area about 260 million years ago. When the sand was buried under a shallow ocean and layers of other sediments, the sand became sandstone.

The mule trail winds past dramatic cliffs of sandstone.

Deeper and deeper we go as the mules pick their way down the trail. Soon, the canyon walls change color again. A new layer surrounds us. It is as tall as a thirty-story building! The cliffs are bright red. The guide calls it Redwall Limestone.

The guide points out more fossils. The fossils formed from animals that lived in an ocean. This area was another ocean full of life 335 million years ago. It seems that oceans came and went several times in this area over the years.

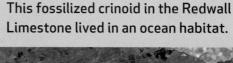
This fossilized crinoid in the Redwall Limestone lived in an ocean habitat.

Later we come to a different layer. The rock is shale. It formed from mud more than 530 million years ago. We also see more sandstone. The shale has smaller **grains** than the sandstone. The small grains make the shale feel smooth.

TECHTREK
myNGconnect.com

Digital Library

ROCK LAYERS OF THE GRAND CANYON

Limestone
Sandstone
Redwall Limestone
Shale
Sandstone
Vishnu Schist

The layers in this close-up of the Bright Angel Shale Formation are easy to find. They show millions of years of history in the Colorado River area.

grain
Grains are small mineral or rock pieces.

Now we are very deep inside the canyon. The walls change color again. They also change texture. The layers are gone! The walls are made of black and white **metamorphic rock** called schist. This rock is nearly two billion years old! Long ago, it was a layer of sedimentary rock. It was squeezed by forces deep underground. The great pressure changed its properties. The layer became schist.

There's something odd about the schist. Here and there you see bands of another rock running through it. That's granite, an **igneous rock** . It formed when magma pushed its way through cracks in the schist. This happened deep underground, so the magma cooled slowly and formed granite.

Look for the property of cleavage in this schist. What other properties do you see? Can you identify the lighter-colored bands of granite?

Today the Colorado River is cutting through schist and granite at the bottom of the Grand Canyon. The ridges of pinkish rock running up the hill are made of granite.

The mule trail ends at Phantom Ranch at the bottom of the canyon.

What a trip it has been! In six hours, we have traveled through two billion years of history. We'll camp by the river tonight and climb back up tomorrow. Then we can observe the history in reverse. Once again, we'll see the different periods of history represented by the changing layers of rocks and fossils. It will be another trip to remember.

These mules enjoy the shade and a well-deserved rest.

metamorphic rock
Metamorphic rock is rock that has been changed by heat or pressure.

igneous rock
Igneous rock is rock that forms when melted rock cools and hardens.

SHARE AND COMPARE

Turn and Talk How can you tell what the land was like around the Grand Canyon at different points in history? Form a complete answer to this question together with a partner.

Read Select two pages in this section. Practice reading the pages. Then read them aloud to a partner. Talk about why the pages are interesting.

my SCIENCE notebook

Write Write a conclusion that tells the important ideas about what you have learned about rocks in the Grand Canyon. State what you think is the Big Idea of this section. Share what you wrote with a classmate. Compare your conclusions. Did your classmate recall that the oldest rocks are at the bottom of the canyon?

my SCIENCE notebook

Draw Think about all the different types of rock in the Grand Canyon. In groups of four draw a picture of what you think the land was like when four different rock layers were formed. Combine the drawings with those of your group and make a presentation to the rest of the class.

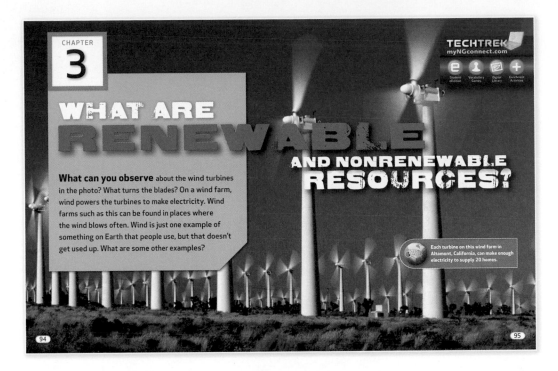

After reading Chapter 3, you will be able to:

- Define and identify natural resources. **WHAT ARE NATURAL RESOURCES?**

- Identify renewable resources. **RENEWABLE RESOURCES**

- Identify nonrenewable resources. **NONRENEWABLE RESOURCES**

- Identify and describe the components and properties of soil. **SOIL**

- Identify ways humans affect Earth's natural resources. **PEOPLE AND RESOURCES**

- Describe ways to help the environment. **PEOPLE AND RESOURCES**

- Science in a Snap! Identify renewable resources. **RENEWABLE RESOURCES**

WHAT ARE RENEW

What can you observe about the wind turbines in the photo? What turns the blades? On a wind farm, wind powers the turbines to make electricity. Wind farms such as this can be found in places where the wind blows often. Wind is just one example of something on Earth that people use, but that doesn't get used up. What are some other examples?

TECHTREK
myNGconnect.com

Student
eEdition

Vocabulary
Games

Digital
Library

Enrichment
Activities

ABLE AND NONRENEWABLE RESOURCES?

Each turbine on this wind farm in Altamont, California, can make enough electricity to supply 20 homes.

SCIENCE VOCABULARY

natural resources
(NA-chur-ul RĒ-sors-es)

Natural resources are materials that are found on Earth that people use. (p. 98)

plants

soil

rocks

water

A forest contains many natural resources.

renewable resources
(rē-NŪ-uh-bul RĒ-sors-es)

Renewable resources are materials that are continually being replaced and will not run out. (p. 98)

People can plant new trees, so trees are a renewable resource.

nonrenewable resources
(non-rē-NŪ-uh-bul RĒ-sors-es)

Nonrenewable resources are materials that cannot be replaced quickly enough to keep from running out. (p. 98)

Coal takes million of years to form so it is a nonrenewable resource.

my Science Vocabulary

conservation
(kon-suhr-VĀ-shun)

fossil fuel
(FOS-ul FYÜ-ul)

natural resources
(NACH-er-uhl RI-sôrs-es)

nonrenewable
(non-ri-NÜ-uh-buhl)

ore
(OR)

renewable
(ri-NÜ-uh-buhl)

TECHTREK
myNGconnect.com

Vocabulary
Games

ore (ÔR)

Ore is rock that contains metal. (p. 108)

The ore in this mine contains copper.

fossil fuel (FOS-ul FYÜ-ul)

A **fossil fuel** is a source of energy that formed from the remains of things that lived millions of years ago. (p. 110)

Fossil fuels, like oil and natural gas, are found between layers of rock.

rock

natural gas
oil

conservation (kon-suhr-VĀ-shun)

Conservation is the protection and care of natural resources. (p. 121)

You can practice conservation by turning off the faucet while brushing your teeth.

Natural Resources

Earth is full of the **natural resources** that people need. Forests are filled with plants, big and small. Water flows over the rocks in creeks. Plants and water are **renewable resources** that are continually being replaced. Water keeps flowing, and new plants sprout. This is not true of resources such as natural gas and coal. They are **nonrenewable resources** . They cannot be replaced quickly enough to keep from running out.

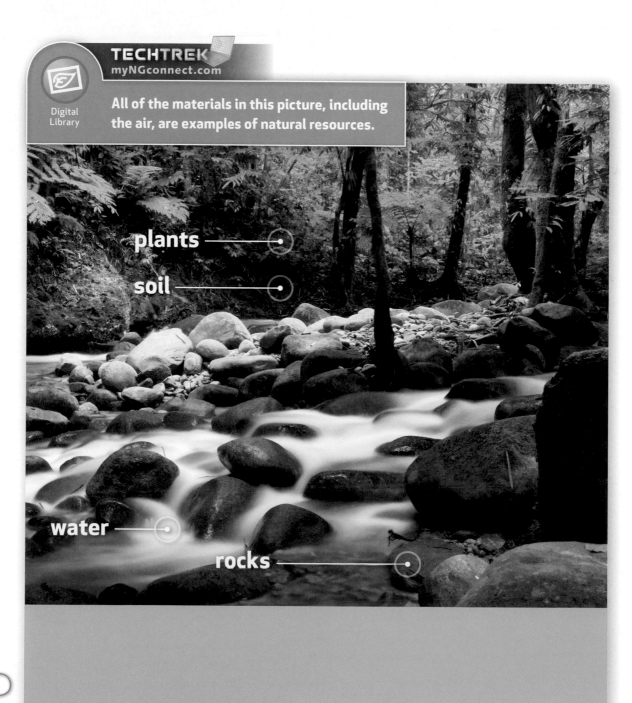

TECHTREK
myNGconnect.com

Digital Library

All of the materials in this picture, including the air, are examples of natural resources.

plants

soil

water

rocks

Use the chart to learn about some of Earth's natural resources.
Which of these resources are renewable and which are nonrenewable?

RENEWABLE AND NONRENEWABLE RESOURCES

Enrichment
Activities

FISH
If people are mindful of how many fish they catch, enough fish will be left to reproduce over and over. So, with careful use, this resource will not run out.

OIL
People use oil at a fast pace. People cannot replace the oil they use.

COAL
Coal formed from plants that died millions of years ago. This resource will run out.

WOOD
New trees can be planted to replace those that people cut down.

Before You Move On

1. Name at least five natural resources.
2. Why are trees considered a renewable resource?
3. **Contrast** How do renewable and nonrenewable resources differ?

Renewable Resources

Living things need water. People could not survive without drinking it.
Plants and animals that people rely on also could not live without it.
People use water when they do laundry, wash dishes, bathe, and swim.
How else do you use water?

THE WATER CYCLE

CONDENSATION
The water vapor cools. It condenses into tiny drops of water that form clouds.

EVAPORATION
As the sun's energy heats the water, it evaporates into water vapor and rises into the air.

The same water is used over and over again. Water continually moves from ground to air and back again. Look at the pictures in the diagram. They show the water cycle. What are the steps in the cycle?

PRECIPITATION
Rain falls from the clouds. It sinks into the soil or enters bodies of water.

THE WATER CYCLE
The cycle repeats itself, renewing the water on Earth.

101

Living Things Have you seen tiny trees sprout in a lawn? Or new plants cover a vacant lot? Plants are a renewable resource. In a forest, seeds drop from full-grown trees and sprout into new little trees. If trees are harvested for sale, new trees can be planted to replace those that are cut down. Harvested trees are sold to make boards, plywood, and posts. Wood pulp from trees is used to make paper.

Seedlings like these can be planted to renew forests.

When trees at this tree farm in Oregon are cut down, farmers plant new ones to replace them.

Plants other than trees can be renewable resources, too. Farmers grow many types of crops, including vegetables to eat, wheat to be ground into flour, and cotton for making cloth. Fields are usually replanted each growing season. Seeds come from plants grown the year before.

Plants also help control erosion. Their roots hold soil together and keep it from blowing or washing away. During floods, plants help limit damage by slowing the movement of water.

The crops growing on this farmland are a renewable resource.

Sunlight and Air Not all natural resources can be seen or touched. Sunlight and air are resources that all living things need. Earth has a constant supply of these resources. They are renewable because they do not run out. The sun is a powerful energy source for the whole world. Living things depend on its energy for heat and light. Earth's weather and the water cycle are powered by energy from the sun.

Science in a Snap! Sun Power

Put a small amount of water in a zip-top plastic bag. Add 10 drops of food coloring to the water. Seal the bag, but leave some air at the top.

Set the bag in the sun. Check the bag later in the day.

What happened to the water? How is your experiment like the water cycle?

Air is one of our most important renewable resources. Humans and most other living things cannot survive without it. Air is a mixture of different gases, including nitrogen, oxygen, and carbon dioxide. When you breathe in, you use oxygen from the air, and when you exhale, you give off carbon dioxide.

You cannot see air, but you can see the effect of moving air, or wind.

Before You Move On

1. List some products that are made from trees.
2. Explain how water is a renewable resource.
3. **Apply** Name two renewable resources you have made use of today and tell how you used them.

Nonrenewable Resources

Rocks Remember that nonrenewable resources cannot be replaced quickly. Examples include rock resources. Rocks form over an extremely long period of time. People cannot simply "grow" a new supply. Rocks are used in many different ways.

A factory worker marks a slab of granite to be cut. This granite will be used as a kitchen countertop.

Large pieces of rocks such as granite and marble are cut and polished to make countertops. Small pieces of rock called gravel are used for roads. Limestone is a key ingredient in cement.

Another rock resource, sand, is melted in a hot furnace to make glass. The hot molten glass can be rolled into sheets to make windows. Molten glass can also be shaped to make bottles, light bulbs, lenses, bowls, and vases.

Cement is mixed with rocks and water to make concrete.

This artist forms glass into a vase. Sand is used to make glass.

Metals Ore is rock that contains metal. The metal is *extracted,* or separated, from the ore. It then goes through a process that turns it into a usable material.

One useful metal is copper. This reddish-orange metal can be melted and shaped. Some copper is pulled into long wires and used in electronic equipment. What can you observe about copper by studying the photos on these pages? What steps does copper go through before it becomes a product?

1. At this mine in Utah, copper ore is loaded into trucks. It is taken to a mill.

2. A furnace heats the ore to a very high temperature. The copper melts and separates from the waste rock.

Other metals are common in everyday products. Aluminum is a lightweight, silver-colored metal. It comes from an ore called bauxite. Aluminum is used to make soda cans, cars, and aluminum foil.

Iron is found in several kinds of ores. Iron is a main ingredient in steel. Steel is used in thousands of products, from paper clips to bridges.

ALUMINUM
Most aluminum soda cans open with pull tabs, which are also made of aluminum.

STEEL
Like many bridges, the Golden Gate Bridge in California is made of steel.

3. The melted copper is cooled into solid metal. The copper is ready to be shaped and cut into products for homes and businesses.

4. Some copper is used to make wires. Copper wires carry electricity around the country. They are used in computers and other equipment.

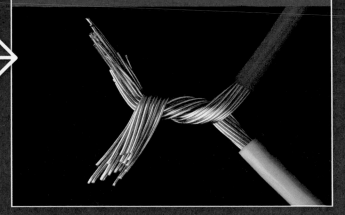

Fossil Fuels Fossil fuels formed from the remains of plants and animals that lived millions of years ago. Coal is a fossil fuel used throughout the world. It formed from trees and giant ferns that died and fell into swamps. Mud piled on top, and the plant remains turned into a spongy material called peat. Over time, the peat was buried even deeper. As air and water were squeezed out, it slowly hardened into coal.

This barge is carrying coal. The coal will be used to help produce electricity at a power plant. Earth's supply of coal is being used up.

HOW A **COAL-FIRED** POWER PLANT WORKS

1. Coal is burned to heat water and make steam.

2. The steam spins the blades of a turbine.

3. The spinning turbine runs a machine called a generator. The generator makes electricity for homes and businesses.

Natural gas and oil are two other fossil fuels that are a valuable source of energy. But like coal, they are running out. Natural gas is burned to heat water and warm homes. Oil is made into fuels that are burned in cars, trucks, and planes. Oil provides a lot more than fuel, however. The asphalt that covers roads and parking lots comes from oil. So do plastics and some paints. Nylon, fleece, and other clothing fabrics come from oil. Even your lip balm is an oil product.

How is oil being put to use at this construction site?

Fuel-powered truck

Nylon vest

Asphalt road surface

Oil and natural gas are found underground between layers of rock. Special drilling equipment is used to reach these resources. Sometimes the rock is beneath the ocean floor. Huge oil rigs are built in the water. Cables attach the rigs to the ocean floor. People live and work on these rigs for weeks at a time to drill through the underwater rock. Special drilling equipment is used to reach these resources.

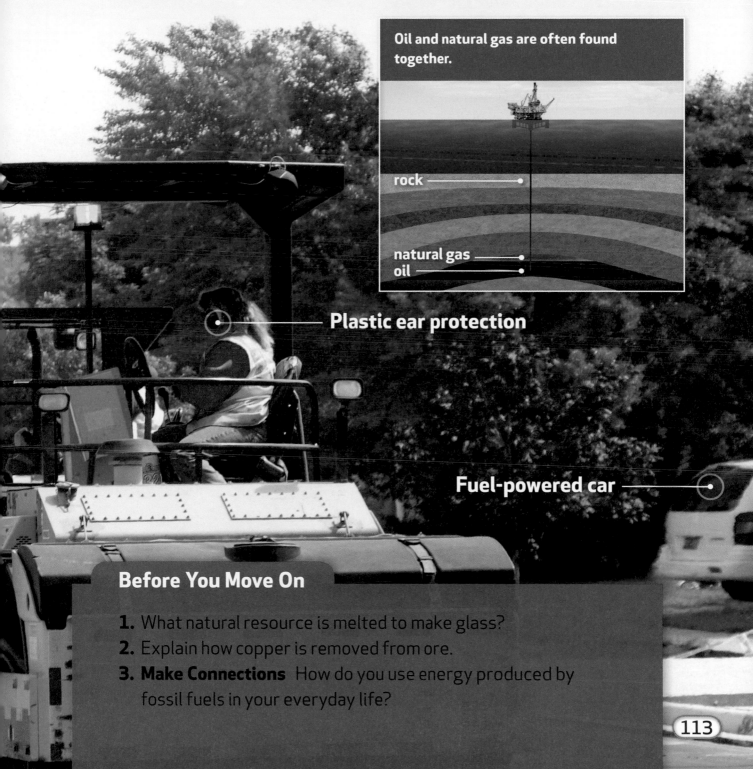

Oil and natural gas are often found together.

rock

natural gas
oil

Plastic ear protection

Fuel-powered car

Before You Move On

1. What natural resource is melted to make glass?
2. Explain how copper is removed from ore.
3. **Make Connections** How do you use energy produced by fossil fuels in your everyday life?

Soil

One natural resource lies right beneath your feet—soil. Soil is found on farmland, grazing land, and forested land. Did you ever look closely at a handful of soil? It is made up of different materials, including tiny particles of rock, water, and air. Our soil resources are nonrenewable because soil forms very slowly over many, many years.

Plants

Plants grow well in this top layer of dark, humus-rich soil.

Plant roots

Rocks

Soil formation begins with weathering. Rain, wind, and ice slowly weather rock, breaking it into tiny pieces that will become part of the new soil. Fallen leaves, dead plants, and dead animals also become part of the soil. Worms and bacteria help break down and decay the dead material, forming a material called humus. The humus and rock particles mix with air and water to make soil.

HOW **SOIL** FORMS

Weathering slowly breaks rock down into tiny particles.

\+

Plant and animal material decays and forms humus.

\+

Animals leave holes that fill with air and water.

Soil is a resource because living things grow in it. Farmers need it to grow potatoes, corn, hay, and other crops. Some of these crops are eaten by people, and some are eaten by animals that are raised for food. Crops, trees, grass, and most of the other plants on Earth need soil to grow. These plants release oxygen into the air. People and other animals need oxygen to breathe.

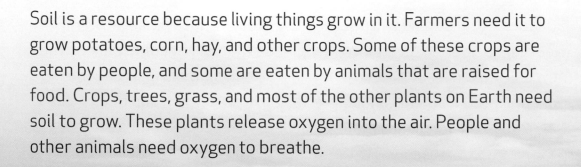

Crops of sugarcane are grown in warm areas of the world. Sugarcane can grow well in sandy soil, clay soil, or loam. Sugar is made from juice found in the tall stalks.

Not all soil is the same. Read the chart to find out about some different kinds of soil. What kinds of soil have you seen where you live?

TYPES OF SOIL

TECHTREK
myNGconnect.com

Digital Library

SANDY SOIL

This soil contains a lot of sand. It absorbs, or takes in, water quickly. It also dries out quickly. Some crops can grow well in it.

CLAY SOIL

This soil contains clay. It absorbs water slowly. Sometimes moving water runs off before it can be absorbed. But clay soil holds the water it absorbs longer than other soils.

LOAM

This soil contains a lot of humus. Loam absorbs water, but not too quickly or too slowly. Water drains through slowly enough for roots to take in the water. It has nutrients and is a good soil for many crops and other plants.

Before You Move On

1. What is soil made of?
2. Explain how plants that grow in soil are related to breathing.
3. **Apply** Sam wants to plant some tomatoes. He knows they need moisture and rich soil. Should he plant them in sandy soil, clay soil, or loam? Explain.

People and Resources

People cannot survive without resources such as air, water, and food. People's demand for resources can affect the environment. For example, people may cut down forests in order to grow more crops. If nonrenewable resources get used up, they are gone. Even renewable resources need to be used carefully.

TECHTREK
myNGconnect.com

Digital Library

Compost enriches soil. To make compost, put grass clippings and leaves in a bin or pile outside. Add fruit and vegetable scraps, too. Each week, stir the pile with a rake. Observe what happens over the coming months. Put your compost on a garden.

Everyone can help prevent damage to our valuable resources. For example, some farmers use no-till farming to help prevent soil erosion. Plowing exposes bare soil that can easily blow or wash away. With no-till farming, farmers do not plow each year before planting. Instead, they leave last season's plant litter in place and plant new seeds right into it.

This plow turns over the soil and exposes it to erosion.

This farmer uses no-till farming to prevent erosion.

Some parts of the world have a larger water supply than others. All places might experience water shortages from time to time, however. Part of the problem is wasted and polluted water. At home, people waste water by running the washing machine when it is not full or by watering the lawn unnecessarily. On farms, poor irrigation methods may waste water. Polluted water usually cannot be used for anything.

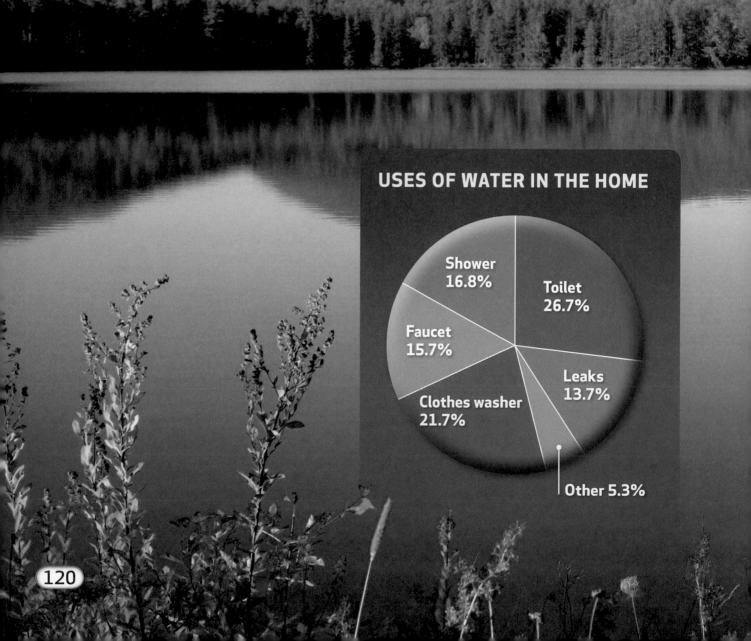

USES OF WATER IN THE HOME

Shower 16.8%

Toilet 26.7%

Faucet 15.7%

Leaks 13.7%

Clothes washer 21.7%

Other 5.3%

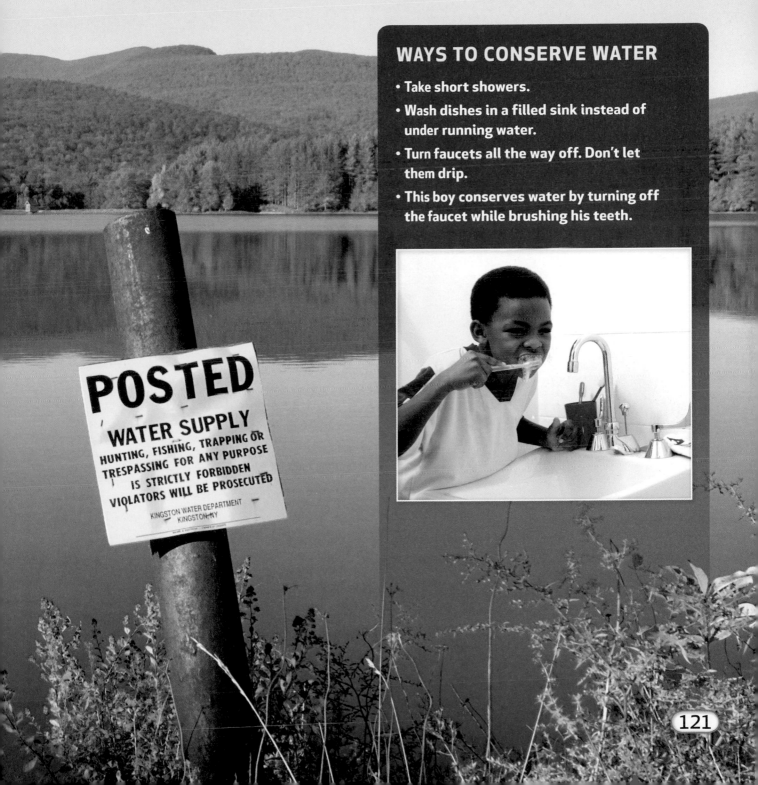

By protecting and caring for water resources, people can make sure there is enough water to go around. The protection and care of a resource is known as **conservation**. People can use washing machines, dishwashers, and other appliances that use less water. Running appliances less often is another way to conserve water. What things can you do to conserve water?

WAYS TO CONSERVE WATER

- Take short showers.
- Wash dishes in a filled sink instead of under running water.
- Turn faucets all the way off. Don't let them drip.
- This boy conserves water by turning off the faucet while brushing his teeth.

POSTED

WATER SUPPLY

HUNTING, FISHING, TRAPPING OR TRESPASSING FOR ANY PURPOSE IS STRICTLY FORBIDDEN VIOLATORS WILL BE PROSECUTED

KINGSTON WATER DEPARTMENT
KINGSTON NY

Many of the products you use come from rock and mineral resources. For example, metal for making electronics, vehicles, and turbines comes from mineral ores. But these resources are getting used up, and high-quality ores are becoming harder to find.

 Bauxite from this mine in Cape York Peninsula, Australia is used to make aluminum.

If people use fewer rock and mineral resources, Earth's supply will last longer. To use fewer metal ores, people can buy fewer cars, electronics, and appliances and replace these items less often. Recycling helps, too. Glass and many metals can be recycled. Aluminum soda cans can be recycled over and over without affecting the quality of the aluminum. Copper and other metals can be melted down and made into new products.

RECYCLED ITEMS	ROCK OR MINERAL RESOURCES
GLASS BOTTLES →	SAND
SODA CANS →	ALUMINUM
FOOD CANS →	IRON, TIN
SCRAP METAL →	IRON, COPPER, ALUMINUM

Recycling helps conserve resources.

Fossil fuels are an important source of energy, but they can cause air pollution. Cars, trucks, and planes use fossil fuels that release polluting gases as they burn. Smoke from coal-fired power stations and industries pollutes the air. Polluted air causes breathing problems for some people.

Polluted air, or smog, hangs over cities where a lot of fossil fuels are burned.

These solar panels collect energy from the sun. Using solar power can help reduce air pollution.

People can reduce air pollution from fossil fuels by reducing our use of those fuels. For example, people can drive more fuel-efficient cars. They can drive less and walk or bicycle more.

Another way to reduce air pollution is by developing cleaner alternative energy sources—wind, sun, heat from the ground, and even moving ocean water. Many exciting ideas are being studied and tried.

More and more people are riding their bicycles to work. Riding bicycles instead of driving cars helps cut down on fossil fuel use.

Can turning off electric lights reduce air pollution? It does if a coal-fired power plant is producing that electricity. Smoke from burning coal pollutes the air.

Before You Move On

1. Give an example of water conservation.
2. How does no-till farming help prevent soil erosion?
3. **Evaluate** Wind and sun energy reduce air pollution. Name another benefit they have over fossil fuels.

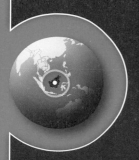

NATIONAL GEOGRAPHIC

RESOURCES IN THE BORNEO RAIN FOREST

Few places in the world are so rich in resources as the rain forest of Borneo. Borneo is an island slightly larger than Texas. It lies on the equator south of China.

Trees and plants in Borneo's rain forest give shelter to many animals and produce oxygen for animals and humans to breathe. Some plants may have materials to make medicines and other products. The rain forest is home to an amazing variety of birds, insects, and other animals, many of which cannot be found in other places.

Orangutans live in the trees of the Borneo rain forest.

Living things are not the only resources found on Borneo. Minerals and fossil fuels are hidden underground. The land itself is valuable, as people have cleared areas of the rain forest to plant rows of oil palms. These trees produce an oil that can be sold for a lot of money. Using the land and digging for resources hurts the island's living resources. If people change the land too much, the plants and animals found there may be lost forever.

Can the remaining rain forest be saved? There is hope. Many people are working to do just that. People who live in Borneo are working to use less of the island's resources. They hope to find ways to support their communities without hurting the beautiful land around them.

Areas of rain forest were cleared for logging and to clear space for crops and roads.

Oil palm plantations take the place of plants and animals that naturally live and grow in Borneo.

Conclusion

Earth has all the resources people need to survive. Renewable resources, such as trees and fresh water, will not run out if used carefully. Nonrenewable resources will run out. Ores, fossil fuels, and soil cannot be replaced. To help current supplies last, people can use less and recycle. Wind power and other alternatives can begin to replace fossil fuels.

Big Idea Natural resources are either renewable, and easily replaced as they are used, or nonrenewable, and not easily replaced because they take too long to form.

Renewable resources + Nonrenewable resources = **NATURAL RESOURCES**

Vocabulary Review

Match the following terms with the correct definition.

A. **fossil fuel**

B. **natural resources**

C. **nonrenewable resources**

D. **ore**

E. **renewable resources**

F. **conservation**

1. Materials that are found on Earth that people use
2. Rock that contains metal
3. Fuels that formed from the remains of things that lived millions of years ago
4. The protection and care of natural resources
5. Resources that are continually being replaced and will not run out
6. Resources that cannot be replaced quickly enough to keep from running out

Big Idea Review

1. **Restate** In your own words, tell what natural resources are.

2. **Describe** How do ores differ from other rocks?

3. **Explain** What role does the sun play in the water cycle?

4. **Sequence** Explain the process of how coal formed.

5. **Analyze** How would driving highly fuel-efficient cars affect air quality? Explain.

6. **Draw Conclusions** Do you think humans could survive without soil? Explain.

my
SCIENCE
notebook

Write About Resources

Make Judgments Which natural resources were used to make soda cans, paper, and glass? Would you encourage people to recycle? Why or why not?

CHAPTER 3 EARTH SCIENCE EXPERT: NATURALIST

Do you love nature?
How about being a naturalist?

Are you curious about the things in nature? Do you care about protecting Earth? You might want to be a naturalist! You could work in a park, greenhouse, or lab. Naturalist Johari Cole works on a farm.

TECHTREK
myNGconnect.com

Digital Library

Johari Cole helps people learn from nature. "Not all teachers are in classrooms—some spin webs."

TECHTREK
myNGconnect.com

Student
eEdition

Digital
Library

As a naturalist, what do you do?

I operate a farm where we take good care of the land and grow healthy food. But I'm also a teacher, guide, researcher, and scientist.

What kind of farm is it?

We raise organic produce on our farm, Being "organic" means we take care of the soil and don't use chemicals to feed the plants or kill pests.

What is a day like on your farm?

Each morning we feed the animals and get ready for any tours or visitors. We plant, water, weed, work in the greenhouse, do research, and sell our produce. We take care of the equipment and computers. And we're always repairing or building new barns and sheds.

What do you like about your job?

Being creative! I like to plan ways to share what we've learned. I want visitors to leave with good memories. They might even get ideas to try at home. The best thing is to see children smile when they eat food that they grew themselves.

How did you prepare to be a naturalist?

I grew up with science books. I found that I loved science, so I studied science in school and college.

Why do you like science?

It never gets boring. There is always something new to discover and explore!

How is Earth science important in your work?

Awareness of Earth and its natural resources is at the root of all the things I do. Honor and respect for Earth guides my decisions and actions.

Johari Cole's Earth-friendly farm south of Chicago attracts many visitors.

BECOME AN EXPERT

Making Blue Jeans:
Resources on the Move

Think of all the things you use every day. Think of all the food you eat. Think of all the clothes you wear. All the things people eat and use come from **natural resources**.

BUTTONS
Many plastics are made using oil.

SHOES
Rubber is made from the sap of a rubber plant.

BELT
Leather comes from animals, and metals are made from ores.

natural resources
Natural resources are materials that are found on Earth that people use.

TECHTREK
myNGconnect.com

Student
eEdition

Digital
Library

Do you have a pair of jeans? Most people do. In fact, people in almost every country wear jeans. Men, women, girls, and boys wear jeans. Where do jeans come from? Jeans are made of denim. Denim is a type of cloth made from a plant called cotton. Cotton plants are natural resources. It takes many steps and natural resources to change cotton into jeans.

Cotton from this cotton field will be used to make blue jeans.

T-SHIRTS
T-shirts are made from the cotton plant.

JEANS
Blue jeans are made from the cotton plant.

From Seed to Cotton

Farmers plant cotton seeds in the soil in early spring. The seeds grow into plants with flowers. When the flowers drop off, seed pods are left behind. After awhile, the pods open to reveal white cotton fibers.

Cotton plants need sunlight, air, and water. These **renewable resources** are always available, but not always in the amounts needed. Cotton grows best with a lot of sun. Cotton needs moisture while it is growing and dry weather once the pods open.

FROM COTTON POD TO COTTON FIBERS

Closed cotton pod

Open cotton pod

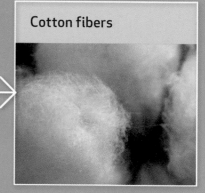

Cotton fibers

This farmer in Guajarat, India, shows some of the crop that will one day become jeans and other clothes.

renewable

Renewable resources are those that are continually being replaced and will not run out.

Picking Cotton Most farmers use large machines to pick the cotton. Next, a machine called a cotton gin takes the seeds out of the cotton fibers. Before cotton gins were invented, people had to separate the seeds from the cotton by hand. Cotton gins speed up the process. Farmers are able to then use the leftover seeds to grow more cotton.

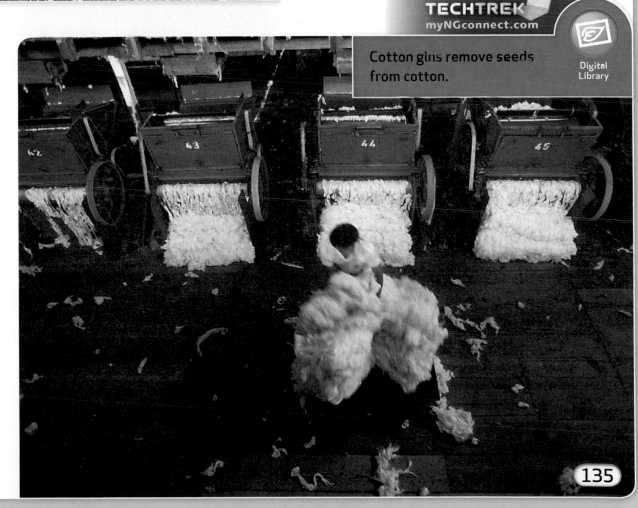

Large machines are used to pick cotton.

TECHTREK
myNGconnect.com

Cotton gins remove seeds from cotton.

Digital Library

After the seeds have been removed from the cotton, a different machine packs the fibers into bales, or large bundles.

Machines play a big role in growing and picking cotton. Energy is needed to run all these machines, including the trucks that carry the cotton from place to place. Currently almost all of this energy comes from **fossil fuels**, but alternatives to these kinds of fuels are being developed.

Trucks usually carry cotton bales to mills. There the cotton becomes cloth.

fossil fuel

A **fossil fuel** is a source of energy, such as coal, petroleum, or natural gas, which formed from the remains of things that lived millions of years ago.

From Cotton to Denim

The bales of cotton are transported to a mill, where leaves, twigs, and dust are removed. Then machines twist and stretch the cotton fibers into thin ropes. Finally, the cotton is spun fast to make it into cotton thread. The cotton is often dyed blue, but some is dyed black or other colors.

At a denim factory, the cotton is woven on a loom to make denim cloth. It is woven with the threads close together to make the denim thick and strong.

A worker checks bales of cotton at a factory.

TECHTREK
myNGconnect.com

Cotton is spun into cotton yarn.

Digital Library

From Denim to Jeans

At the factory, people use a special machine to cut the denim into pieces to make the jeans. Other people use sewing machines to sew the pieces together. Thick thread is used for sewing. Jeans are made to be strong.

Some jean parts are not made from cotton. The zippers and buttons are steel. Copper rivets are attached at the corners of the pockets. The metal to make these parts comes from underground **ores** . Ores are rocks that contain metals. They are **nonrenewable resources** .

Jean zippers are often made from steel.

A woman sews jeans at a factory.

ore

Ore is rock that contains metal.

nonrenewable resource

Nonrenewable resources are those that cannot be replaced quickly enough to keep from running out.

Boxes of jeans are sent from the factory to warehouses. Stores order jeans from the nearest warehouses, and trucks deliver them.

Finally the jeans are put onto the store shelves. Will you buy a pair? If you do, remember how many resources it takes to make jeans. Don't let the resources go to waste.

Practice **conservation** by using your jeans until they wear out. If they get too small for you, give them away to someone who can use them. Everyone likes a pair of jeans!

Jeans are put onto store shelves so people can buy them.

conservation
Conservation is the protection and care of natural resources.

CHAPTER 3
SHARE AND COMPARE

Turn and Talk How does part of a cotton plant in India end up in your closet? Form a complete answer to this question together with a partner.

Read Select two pages in this section. Practice reading the pages. Then read them aloud to a partner. Talk about why the pages are interesting.

my SCIENCE notebook **Write** Write a conclusion that tells the important ideas you have learned about the use of natural resources in making blue jeans. State what you think is the Big Idea of this section. Share what you wrote with a classmate. Compare your conclusions. Did your classmate recall that ores are used?

my SCIENCE notebook **Draw** Draw a step in the process of making a pair of blue jeans. Label it with the kinds of natural resources that are used in that step. Put your drawings together with your classmates in "field to closet" order.

HOW DO SLOW PROCESSES CHANGE EARTH'S SURFACE?

All over Earth you can see many natural features, such as mountains, valleys, canyons, and cliffs. You might think that these features never change, but they do. Slowly over time, mountains wear away, and valleys widen. Canyons deepen, cliffs crumble to the ground, and new land forms. Earth's surface is changing all the time.

TECHTREK
myNGconnect.com

Water rushes over rocks in a lake fed by water from Vatnajokull Glacier in Iceland.

142   143

After reading Chapter 4, you will be able to:

- Describe that Earth's surface can change slowly. **WEATHERING, EROSION AND DEPOSITION**

- Identify the major landforms on Earth's surface. **LANDFORMS ON EARTH'S SURFACE**

- Identify and describe how weathering changes Earth's surface. **WEATHERING, CAUSES OF WEATHERING**

- Identify and describe how erosion and deposition change Earth's surface. **EROSION AND DEPOSITION, CAUSES OF EROSION AND DEPOSITION**

- Identify how weathering, erosion, and deposition affect people. **WEATHERING AND EROSION AFFECT PEOPLE**

- Science in a Snap! Identify and describe how weathering changes Earth's surface. **CAUSES OF WEATHERING**

HOW DO SLOW PROCESSES CHAN

All over Earth you can see many natural features, such as mountains, valleys, canyons, and cliffs. You might think that these features never change, but they do. Slowly over time, mountains wear away, and valleys widen. Canyons deepen, cliffs crumble to the ground, and new land forms. Earth's surface is changing all the time.

TECHTREK
myNGconnect.com

Student eEdition

Vocabulary Games

Digital Library

Enrichment Activities

GE EARTH'S SURFACE?

Water rushes over rocks in a lake fed by water from Vatnajokull Glacier in Iceland.

SCIENCE VOCABULARY

landform (LAND-form)

A **landform** is a natural feature on Earth's surface. (p. 146)

A sand dune is a common landform in many deserts.

weathering (WE-thur-ing)

Weathering is the breaking apart, wearing away, or dissolving of rock. (p. 148)

Waves crash into rocks and make them smaller through weathering.

sediment (SED-ah-mint)

Sediment is material that comes from the weathering of rock. (p. 152)

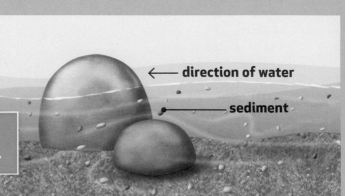

← direction of water

sediment

Sediment in rushing water rubs against the rocks in the riverbed.

my Science Vocabulary

abrasion
(a-BRĀ-zhun)

deposition
(de-pō-ZI-zhun)

erosion
(ē-RŌ-zhun)

landform
(LAND-form)

sediment
(SED-ah-mint)

weathering
(WE-thur-ing)

abrasion (a-BRĀ-zhun)

Abrasion is the scraping away of materials. (p. 152)

Wind blew sediment against the surfaces of these rocks and caused abrasion.

erosion (ē-RŌ-zhun)

Erosion is the picking up and moving of sediment to a new place. (p. 158)

Glaciers move sediment in the process of erosion.

deposition (de-pō-ZI-zhun)

Deposition is the laying down of sediment and rock in a new place. (p. 159)

Deposition formed a fan-shaped delta at the mouth of this river.

delta

Landforms on Earth's Surface

Earth's surface is made up of landforms . A landform is a natural feature. Some landforms, such as mountains and hills, rise high above the surrounding land. Other landforms, such as river valleys and canyons, have higher land around them. Plains are areas of flat land. Deserts are mainly dry, and coastlines are landforms where the land and ocean meet. Use the chart on the next page to learn about some of the landforms on Earth's surface.

Sand dunes stretch across the Sahara, in North Africa. They are landforms on Earth's surface.

LANDFORMS

MOUNTAINS
Mountains are high places with steep sides that rise above surrounding land.

THE HIMALAYA, BHUTAN

PLAINS
A plain is a large flat area of land.

PLAINS IN KANSAS, USA

OCEANS
An ocean is a very large body of salt water.

PACIFIC OCEAN

RIVER VALLEYS
A river valley is low land formed by flowing water.

YELLOWSTONE RIVER, WYOMING, USA

COASTLINES
A coastline is where land meets the ocean.

CANARY ISLANDS, ATLANTIC OCEAN

CANYONS
A canyon is a deep, narrow valley with steep sides.

GRAND CANYON, ARIZONA, USA

Before You Move On

1. Name and describe three landforms you have seen.
2. What is the difference between a valley and a canyon?
3. **Evaluate** On which landform do you think it would be best to build a city?

Weathering

Over millions of years, landforms are destroyed and created. These changes to Earth's surface are part of a process that involves three different actions. One action in this process of change is **weathering**. Weathering is the breaking apart, wearing away, or dissolving of rock. Wind, water, ice, chemicals, and even plants can cause weathering.

 Waves crash against the shore at Big Sur, California. The waves break rocks into smaller rocks through the process of weathering.

Weathering breaks large rocks, such as these cliffs, into smaller rocks.

Water continues to weather smaller rocks. They break into even smaller pieces.

If you look closely at rock, you might see many different colors and shapes. The colors and shapes are minerals that make up the rock. Most rocks are made of one or more minerals. Rock weathers in more than one way. Most of the time when rock is weathered, it gets smaller, but the minerals stay the same. Only the size of the rock changes.

Grains of sand are really just tiny rocks, the product of weathering. This sand is made mainly of a mineral called quartz.

149

Chemicals, such as acids, can also cause weathering. In this type of weathering, the minerals that make up rock actually change. Rain can become a weak acid as it falls through the air. This weak acid seeps into the cracks of rocks, such as limestone. The acid dissolves the minerals in the limestone and the minerals wash away.

These limestone cliffs are located in Yorkshire, England. The large cracks in the limestone were created by water containing acid.

Many caves form in this way. The weak acid can seep underground and dissolve rock. Over time the action of the acid can hollow out tunnels and chambers in the rock. If the water level drops, the chambers fill with air and become caves.

TECHTREK
myNGconnect.com

Digital
Library

HOW A **CAVE FORMS**

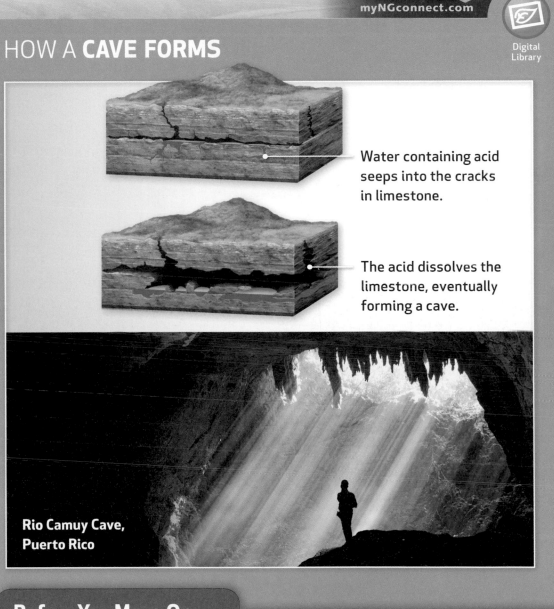

Water containing acid seeps into the cracks in limestone.

The acid dissolves the limestone, eventually forming a cave.

Rio Camuy Cave, Puerto Rico

Before You Move On

1. What happens to rock as it weathers?
2. How does acid weather rocks?
3. **Infer** Look at the photograph on these pages. What do you think will happen to the cracks in the limestone over time?

Causes of Weathering

Weathering by Wind Wind can weather or wear away rock. Wind can pick up small pieces of sediment such as sand. Sediment is the material that comes from the weathering of rock. The wind slams sand against the rock and chips away at it. The rock becomes smooth and rounded, just like a piece of wood that has been rubbed with sandpaper. This type of weathering of a rock's surface is called abrasion .

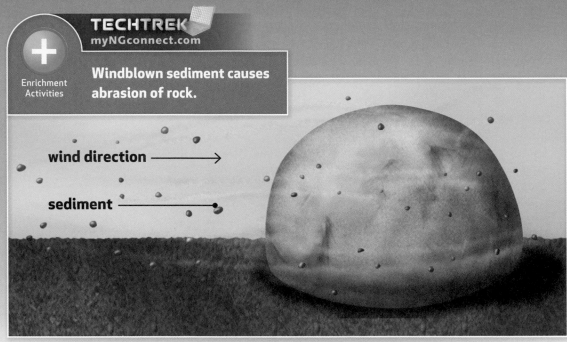

TECHTREK
myNGconnect.com

Enrichment Activities

Windblown sediment causes abrasion of rock.

wind direction

sediment

Wind blows sediment against rocks and wears them away.

Observe a piece of chalk.
Rub it against sandpaper 10 times.
Observe the chalk again.

Repeat with another piece of chalk and construction paper.

What type of weathering does this activity show?
Why did one piece of chalk weather more than the other?

Weathering by Water Have you ever noticed that rocks in a river or at the beach are often smooth and rounded? Just like wind, moving water carries sediment, such as sand and gravel. The sediment in the water constantly bumps and rubs against larger rocks. Eventually, the surface of the rocks wears away. The rocks become smoother, more rounded, and smaller.

The rushing water weathers these rocks in the river.

Sediment in the rushing water causes abrasion, making the rocks smooth.

← direction of water

→ sediment

Moving water can weather rock in a big way. For example, the fast-moving Colorado River formed the Grand Canyon. Sediment, carried along by the water, scraped and chipped away rock on the sides and bottom of the river. It has taken millions of years for the Colorado River to carve the deep canyon. Today the Grand Canyon measures over 1.6 kilometers (1 mile) deep in places. Weathering may be a slow process, but over time, it can make huge changes.

This is the Grand Canyon in Arizona. The Colorado River continues to carve through rock, making the canyon deeper.

Weathering by Plants What do you think happens when the roots of this tree grow into the cracks in the rock? As the tree grows, the roots get bigger. The expanding roots push against the sides of the crack and force the crack to widen. Eventually, the crack gets wide enough to split the rock apart.

Observe how the tree roots grow across the flat rock and into cracks.

Weathering by Ice Ice can weather rock, too. Sometimes water seeps into small cracks in rocks. When the temperature drops below freezing, the water turns to ice. As the water freezes, it expands, or takes up more space. The ice acts like a wedge pushing against the sides of the rock, making the crack wider. Over time, more water seeps into the crack and the process repeats. The repeated freezing and thawing of water can split the rock apart.

What do you think will finally happen to this rock?

Before You Move On

1. What is sediment?
2. Describe abrasion by wind and water.
3. **Compare** How are weathering by plants and ice similar?

Erosion and Deposition

Weathering can create a lot of loose material—boulders, pebbles, and sand. What happens to these pieces of rock? Often, they are moved to a new place. The moving of sediment from one place to another is called **erosion**. Wind, water, ice, and gravity all can move sediment from one place to another, causing erosion.

Think how powerful the glacier must have been to move these large boulders on Deer Isle, Maine.

What happens to the sediment that is carried away? It is deposited, or dropped in a new place, in a process called **deposition** . The soil that erodes from a hill might be deposited at the bottom of the hill. The picture on these pages shows that erosion and deposition happen to sediment of all sizes. Long ago, huge masses of ice called glaciers carried these boulders from far away. When the glacier melted, the boulders dropped in this new place.

Before You Move On

1. Name one way you can tell that erosion is taking place.
2. What is deposition?
3. **Compare and Contrast** How are erosion and deposition alike and different?

Causes of Erosion and Deposition

Erosion and Deposition by Wind Wind erosion also changes the land. Most wind, however, can pick up only tiny particles, such as dust. But the wind can carry that dust amazing distances. The color of this daytime sky in Greece was caused by dust. The dust came all the way from Africa! Strong winds whip up clouds of dust in deserts. Africa's Sahara is the largest desert on Earth. It has huge sandstorms that sometimes lift dust 4.5 kilometers (almost 3 miles) up in the air.

Dust from the Sahara in Africa made it hard to see in this town in Greece.

The dust cloud is sometimes so big it can even be seen from space! Observe in the photo below that a giant dust cloud has blown out over the Atlantic Ocean. The dust can keep moving as long as the wind blows. What happens when the wind stops blowing? As the wind dies down, gravity takes over. The dust drops to Earth's surface. This is an example of wind deposition.

Atlantic Ocean

Africa

This photo was taken from space. The brown swirl is dust. The white patches are clouds.

Erosion and Deposition by Water Does the color of the river in the picture seem unusual to you? The brown color of the water comes from the soil and other sediment being carried by the river. The water is moving the sediment from one place to another. This is an example of erosion by water. Ocean water also moves sediment. Waves pick up sand and move it along the coast.

Gravity is moving the water. Why do you think the water moves faster in some places and slower in others?

Even gentle waves erode some grains of sand.

Rivers and oceans can move large amounts of sediment. What happens when moving water slows down? As water slows down, it deposits the sediment it has been carrying. For example, rivers slow down as they flow into an ocean. The slowed river water deposits sediment in the ocean. Over time, the sediment builds up to form new land called a delta.

This delta has a fan shape. As the river crosses the delta it splits into many channels before it reaches the ocean.

delta

Erosion and Deposition by Ice

What can move rocks as big as a house and as small as a grain of sand? Would you believe that ice can? Ice is the most powerful agent of erosion. Huge glaciers of ice slowly move over the land. As a glacier moves, it plucks rocks from the ground and carries them along.

TECHTREK
myNGconnect.com

Digital Library

Glaciers pick up and move rocks of different sizes.

The Dawes Glacier in Alaska is being pulled downhill by gravity. The stripes are sediment.

Glaciers can carry an enormous amount of sediment. Over time, glaciers melt, or retreat. As the ice melts, deposition occurs. The sediment carried by the glacier is left behind. This deposited sediment is called till. Sometimes glaciers leave behind mounds of till, called moraines. Moraines can form at the front of a glacier or along each side.

A moraine formed from sediment that was pushed in front of a glacier in New Zealand.

moraine

Before You Move On

1. Describe wind erosion and deposition.
2. Which do you think can move bigger sediments, a fast-moving river or a slow-moving river?
3. **Compare and Contrast** How are erosion and deposition by a river like erosion and deposition by a glacier?

Weathering and Erosion Affect People

Weathering, erosion, and deposition are slow, steady processes. But they can cause sudden changes. These changes can be dangerous for people living nearby. For example, a sinkhole forms when a roof of a cave falls in, leaving a hole or pit in the ground. Sinkholes occur in areas where the main rock under the soil is limestone. Over time, caves form in the limestone underground. If the roof of the cave falls in, whatever is on top of the ground falls in too. How do you think the caves formed underground?

Buildings, cars, and trees collapsed into this sinkhole in Winter Park, Florida. Thousands of sinkholes dot Florida. Some fill with water and become ponds. Others form under ponds and drain them!

Weathering and erosion can team up and send rock and soil sliding downhill. That's a landslide. Over time, weathering loosens rock on a slope. Then gravity sends the rock tumbling down. Rain can also cause a landslide. Water loosens the soil and it slides down as mud. Sometimes heavy rains cause rivers of mud that flow quickly downhill. A large landslide can knock houses off their foundations and topple trees. Some have even buried parts of towns.

Parts of this steep hillside gave way and flowed into this community in La Conchita, California.

Mud rushed down this mountainside in the Philippines and buried an entire village.

Changing Land If you want to see land on the move, head to the coast. Waves constantly pound the shoreline, blasting away rock bit by bit. Through weathering and erosion, waves break and move rock. Cliffs change shape as rocks tumble into the sea.

Ocean currents also carry sediment along the coast. Waves deposit sand onshore as beaches, then wash it away and deposit it someplace else.

Storm waves swept away the sand under and around this house. Buildings on the shore are always in danger as the land beneath them shifts and changes.

What may happen to these houses as the cliffs continue to weather and erode?

People try to slow or change the effects of erosion and deposition along coasts. For example, walls of rock and concrete along the shore help protect property from the force of breaking waves. Some people build walls that stick out into the water. The wall traps sand and builds up a beach on one side. But this takes sand away from the other side of the wall, where the beach erodes faster. Some communities truck in tons of sand to replace eroded beaches. But nothing can stop erosion and deposition.

This pile of rocks, called a jetty, helps protect houses and businesses from erosion.

Helpful Effects of Erosion and Deposition Weathering, erosion, and deposition also provide one of the most helpful resources we have—soil. Soil is mostly weathered rock mixed with air, water, and decayed bits of plants and animals.

Most soil did not form in the place you find it. Instead, it eroded from one place and was deposited in another. For example the rich soils of the midwestern United States were deposited when glaciers melted long ago.

The rich soil of the Midwest region of the United States makes excellent farmland, such as the soil on this farm in Indiana.

Farmers have grown crops in the rich soil of the Nile River floodplain in Egypt for thousands of years.

Some of the best soils in the world are found in wide, flat river valleys. These areas are floodplains. They are dry most of the time. But during floods, rivers overflow their banks. A thin layer of water spreads over the land. As the flow of water slows down, the river deposits soil that has eroded off the land farther up the river. This eroded soil is rich in nutrients. While floods can do a lot of damage, the rich soil they leave behind has helped produce food around the world.

This painting shows an ancient Egyptian farmer harvesting wheat.

Before You Move On

1. What causes landslides?
2. How do floodplains help people?
3. **Explain** How do weathering and erosion affect people?

NATIONAL GEOGRAPHIC

SHRINKING GLACIERS
IN SOUTH AMERICA

The Tuni Condoriri is a mountain range in a country called Bolivia in South America. Snow covers the peaks, and glaciers fill the valleys—at least they used to. Some of the glaciers remain, but they are shrinking. In 2009, the Chacaltaya Glacier disappeared. Another nearby glacier may have only 30 years left.

2003

2009

The glacier near the top of Mt. Chacaltaya used to be a ski area. But warmer weather has melted the glacier over the years. Now it is a bare, rocky mountainside.

Many people in Bolivia depend on the glaciers for most of their water. During the summer, some of the glacial ice melts. The water feeds into huge lakes and supplies power plants. Some of the water is used for drinking and watering crops.

The glaciers are certainly important to everyday life in this area. But the glaciers are melting. Why? Earth is getting warmer, in part because of people. We burn fuels such as coal to make electricity. Burning these fuels produces a gas that goes into the atmosphere. This gas acts like a blanket and makes Earth warmer.

People need water for crops and animals.

People need water for many things.

People need water to drink.

These women are walking between their home city of El Alto and the Bolivian capital of La Paz.

Conclusion

Many kinds of landforms make up Earth's surface. They are slowly changed through the processes of weathering, erosion, and deposition. Through weathering, Earth's surface breaks apart, wears away, and dissolves. Then erosion moves the weathered material to a new place. Finally, the rock and other weathered material are laid down by deposition. The actions of water, wind, ice, and gravity may be slow, but effects are great, especially when they happen near where people live and work.

Big Idea Weathering, erosion, and deposition are slow processes that work together to change Earth's surface slowly.

Weathering

+

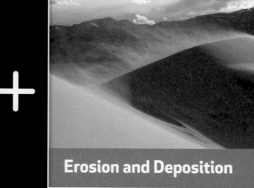

Erosion and Deposition

=

Earth's Changing Surface

Vocabulary Review

Match the following terms with the correct definition.

A. **weathering**   1. The scraping away of materials

B. **sediment**   2. The breaking apart, wearing away, or
                           dissolving of rock
C. **abrasion**
                        3. Material that comes from the weathering of rock
D. **landform**
                        4. The moving of sediment and rock from one
E. **erosion**   place to another

F. **deposition**   5. The laying down of sediment in a new place

                        6. A natural feature on Earth's surface

Big Idea Review

1. **Describe** Name four different kinds of landforms and describe their characteristics.

2. **Restate** How is erosion different from deposition?

3. **Explain** What role does gravity play in erosion and deposition?

4. **Cause and Effect** What changes does weathering make in the size of rocks?

5. **Infer** Suppose you are in a place where temperature often drops below freezing, and you find a rock split in two. What type of weathering could have caused the rock to split?

6. **Evaluate** Consider two hills that are both covered in forest. But one has steep sides and the other one doesn't. On which is a landslide more likely to happen? Why?

Write About Earth's Surface

Draw Conclusions What happened in this photo? Write a paragraph that describes what processes could have led to this result.

EARTH SCIENCE EXPERT: GLACIOLOGIST

What Does a Glaciologist Do?

Have you ever wondered what it's like to work in Antarctica? Its ice provides Lonnie Thompson with prime working conditions. As a glaciologist, Thompson studies glaciers. He drills deep into glaciers and pulls out long tubes of ice. Glaciers do more than erode the land. They teach us about weather and climate. Glaciologists study the bottom of the ice core—the oldest part—to learn what the weather and climate was like a long time ago.

Information from the past can tell scientists about the future. As Thompson says, "Glaciers are an early warning system for the climate of Earth." His study of the ice cores shows that some glaciers are melting quickly. Based on his data, Thompson estimates that the snow on Mount Kilimanjaro—the highest peak in Africa—may be gone by 2020.

TECHTREK
myNGconnect.com

Digital Library

Lonnie Thompson holds a drilled piece of ice core.

Glaciologists use tools such as drills. This drill bit is so sharp it can slice through many meters of solid ice.

TECHTREK
myNGconnect.com

Student
eEdition

Digital
Library

Glaciologists use special tools, such as ice drills, to collect their samples of ice. Glaciologists also measure the growth of glaciers. These scientists climb mountains and lead expeditions to collect their samples. Glaciologists do a lot of hands-on investigation and lead adventurous lives! Thompson has traveled to five continents and has climbed many mountains.

Are you interested in glaciology? Like Thompson, most glaciologists studied science in school. They then earned a degree in geology or glaciological sciences from a university. Glaciologists often work for ice research companies and universities.

Lonnie Thompson's team examines an ice core they drilled from the Coropuna ice cap in Peru, South America.

BECOME AN EXPERT

Yosemite Valley: Shaped by Weathering and Erosion

Yosemite Valley is at the heart of California's Yosemite National Park. It is a broad valley ringed by cliffs and high waterfalls. Most people come to Yosemite for its beautiful scenery, but few realize that it was shaped by glaciers long ago. What clues did the glaciers leave behind? The main clues are the **landforms**, or natural features of Yosemite Valley, including the valley itself. The valley is wide, deep, and U-shaped—a sure sign that glaciers were once there. On either side of Yosemite Valley are high cliffs that were shaped in part by glaciers.

El Capitan

Many of the landforms in Yosemite Valley were formed by glaciers—from the deep valleys to the steep cliffs.

landform

A **landform** is a natural feature on Earth's surface.

TECHTREK
myNGconnect.com

Student
eEdition

Digital
Library

Yosemite
Falls

Half Dome

El Capitan

Merced River

Cathedral
Rocks

Bridalveil
Fall

Cathedral
Rocks

Bridalveil
Fall

Glaciers of Yosemite Valley

Yosemite Valley was occupied by glaciers at least three times over the past million years. Sherwin was the oldest and largest glacier. It filled Yosemite Valley almost to its rim and did most of the work in carving out the valley. The Tahoe Glaciers that filled the valley after the Sherwin Glacier were much smaller. The Tioga Glacier was the most recent and smallest glacier in Yosemite Valley.

WHEN WERE GLACIERS IN YOSEMITE?

Sherwin Glacier —   Tahoe Glaciers —   Tioga Glacier

| 2 MILLION YEARS AGO | 1.5 MILLION | 1 MILLION | 500,000 | 250,000 | TODAY |

HOW THE VALLEY MAY HAVE LOOKED

10 MILLION YEARS AGO

The Merced River cuts the main valley. Notice the mountain that will become Half Dome.

30,000 YEARS AGO

The Tioga Glacier half-fills the valley. Earlier glaciers have sliced the mountain that is Half Dome.

10,000 YEARS AGO

The last glacier has melted. Its moraine has dammed the valley, creating an ancient lake.

How Ice Shaped Yosemite Valley

How was Yosemite shaped by ice? The process began with ice wedging, which is a type of **weathering**.

Ice under a glacier sometimes melts and seeps into cracks in the rock surface. When the temperature drops, the water freezes back into ice. When water freezes, it expands. The ice acts like a wedge and pushes the crack apart. Eventually, the crack gets so big it breaks the rock apart.

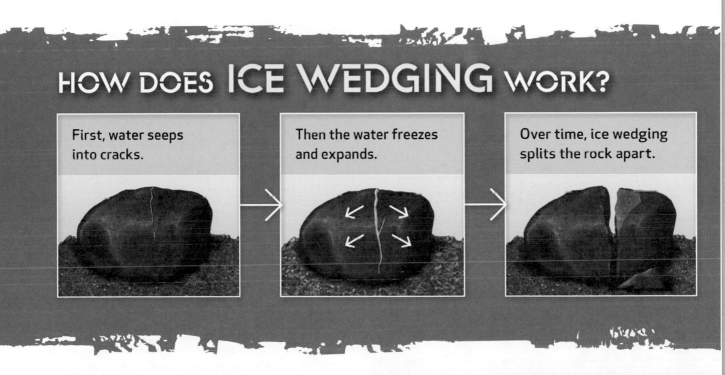

HOW DOES ICE WEDGING WORK?

First, water seeps into cracks.

Then the water freezes and expands.

Over time, ice wedging splits the rock apart.

Ice wedging pried apart a small crack in this rock until it split in half.

weathering
Weathering is the breaking apart, wearing away, or dissolving of rock.

U-shaped Valleys Yosemite Valley looked a lot different before the glaciers arrived. A river cutting through the rock created a fairly small and narrow V-shaped valley. Then the slow-moving but powerful glaciers eroded the valley deeper and wider into the shape of a U. Yosemite doesn't have the rounded valley floor of a typical U-shaped valley. Instead, the floor is flat. That's because a lake formed in the valley after the last glacier melted. Sediment from the surrounding land filled the lake. This sediment covered the rounded floor and created a flat surface. Today, a river on the valley floor is slowly eroding the lake sediments.

Sediments on the valley floor hide the U-shape of Yosemite Valley.

U-shape

Sediment

U-shape Valley

Hanging Valleys After rocks are broken down and loosened from Earth's surface, they are picked up and carried away through the process of **erosion** . Weathering and erosion worked together to create many of Yosemite's landforms.

Bridalveil Fall drops hundreds of meters to the valley floor. The waterfall tells a story of the glaciers that carved out the deep, main valley of Yosemite. Smaller glaciers flowed into the larger glacier, just like small streams joining a big river. The smaller glaciers did not cut as deeply. So when the ice melted, the smaller side valleys were left hanging high above the main valley. In fact, they are called hanging valleys! The rivers that run through these hanging valleys can create towering waterfalls.

HOW DO HANGING VALLEYS FORM?

Side glaciers

Main glacier

Hanging valleys

U-shaped valley

Bridalveil Fall flows from a river in a hanging valley in Yosemite.

erosion
Erosion is the picking up and moving of sediment to a new place.

Glacial Polish The glaciers that carved out Yosemite's valleys carried huge amounts of **sediment**, or weathered rock. Some of this sediment was crushed as it was carried under the glaciers through Yosemite Valley. Imagine thousands of tons of ice crushing the rocks underneath it into a fine powder. As the glaciers moved, this rock powder rubbed against the ground through a process called **abrasion**. In some places, the fine powder rubbed against the rock surface until it became smooth and shiny.

This glacial polish in Yosemite was created long ago through abrasion by glaciers.

sediment
Sediment is material that comes from the weathering of rock.

abrasion
Abrasion is the scraping away of materials.

Moraines When glaciers melt, or retreat, **deposition** begins. The glaciers leave behind loads of sediment, called till, that had been carried in the ice. Till is made up of sediment of all different sizes, from tiny pieces of rock powder to large boulders. When glaciers deposit hills or ridges of till, moraines are formed. Yosemite Valley is covered with moraines that were left as the last glaciers retreated.

HOW DO MORAINES FORM?

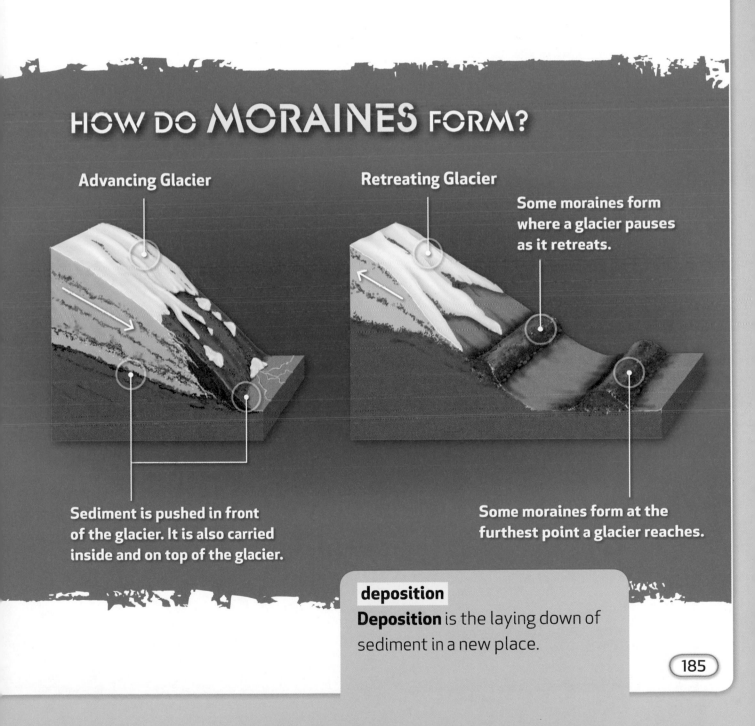

Advancing Glacier

Retreating Glacier

Some moraines form where a glacier pauses as it retreats.

Sediment is pushed in front of the glacier. It is also carried inside and on top of the glacier.

Some moraines form at the furthest point a glacier reaches.

deposition
Deposition is the laying down of sediment in a new place.

Erratics You can find large boulders perched along trails and cliffs throughout Yosemite. Often, these boulders, called erratics, look very different from the rock surrounding them. This is because they were carried a long way by glaciers and deposited in a new place as the ice melted. These boulders are more clues that glaciers once covered Yosemite Valley.

TECHTREK
myNGconnect.com

Digital
Library

This large erratic was carried by a glacier down Yosemite Valley.

Weathering, erosion, and deposition have shaped Yosemite Valley over millions of years. Powerful glaciers plowed through the valley, scraping, gouging, and plucking rocks from the land. The glaciers carried and dumped the rocks in different places. The glaciers are long gone, but the same processes continue to reshape the valley. Wind and rivers weather, erode, and deposit sediment in Yosemite. The landscape is always changing—slowly but surely.

CHAPTER 4

SHARE AND COMPARE

Turn and Talk How did weathering, erosion, and deposition form Yosemite Valley? Form a complete answer to this question together with a partner.

Read Select two pages in this section. Practice reading the pages. Then read them aloud to a partner. Talk about why the pages are interesting.

Write Write a conclusion that tells the important ideas you have learned about how weathering, erosion, and deposition shaped Yosemite Valley. State what you think is the big idea of this section. Share what you wrote with a classmate. Compare your conclusions. Did your classmate recall that ice wedging can crack rocks apart?

Draw Form groups of four. Have each person draw a picture of a landform in Yosemite Valley that was created by a glacier. Add labels to your drawings. Put the drawings together to create a wide-view picture of the valley.

CHAPTER 5

WHAT CHANGES DO VOLCANOES AND EARTHQUAKES CAUSE?

The surface of Earth is always changing. Most of these changes take millions of years. But sometimes an event will change Earth's surface quickly. A mountain can begin to grow in a person's lifetime. An island can rise up out of the sea, or a river can change direction.

TECHTREK
myNGconnect.com

This lava is flowing from Kilauea, Hawai'i's most active volcano. What will happen to the lava as it cools?

190 191

After reading Chapter 5, you will be able to:

- Identify and describe how Earth's surface can change rapidly. **EARTHQUAKES, VOLCANOES, LANDSLIDES**

- Explain that Earth is made up of layers and Earth's surface is divided into plates. **EARTH'S STRUCTURE**

- Explain how earthquakes happen, and can change Earth's surface. **EARTHQUAKES**

- Explain how volcanoes happen, and can change Earth's surface. **VOLCANOES**

- Explain why landslides happen and how they change Earth's surface. **LANDSLIDES**

- **Science in a Snap!** Explain how earthquakes happen, and can change Earth's surface. **EARTHQUAKES**

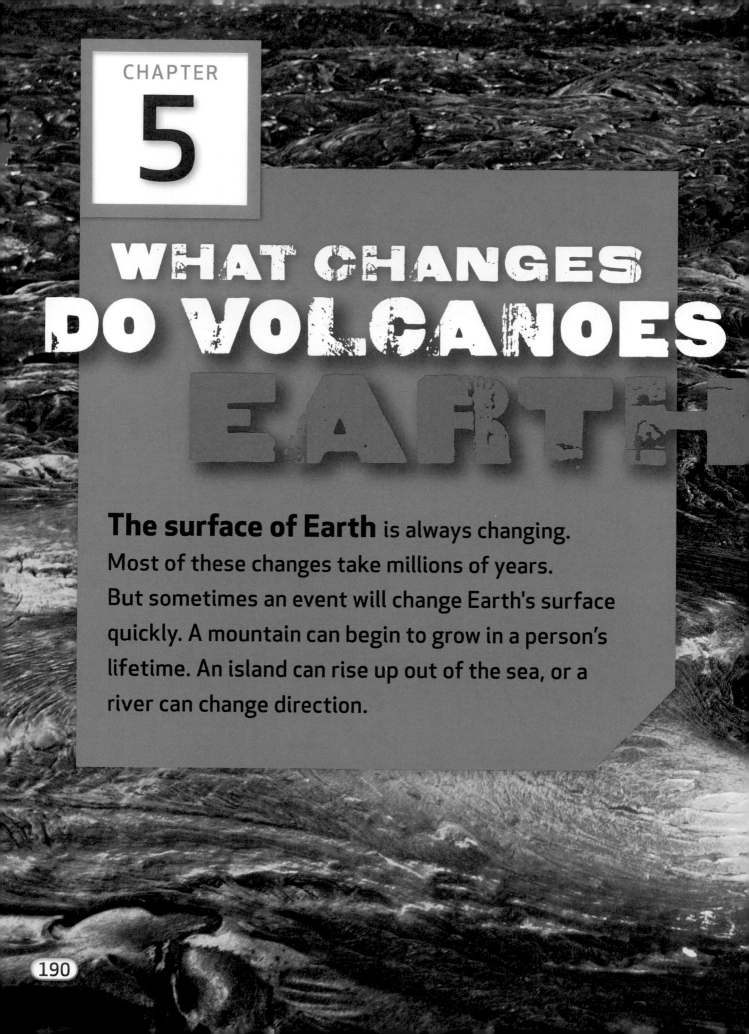

WHAT CHANGES DO VOLCANOES EARTH

The surface of Earth is always changing. Most of these changes take millions of years. But sometimes an event will change Earth's surface quickly. A mountain can begin to grow in a person's lifetime. An island can rise up out of the sea, or a river can change direction.

AND QUAKES CAUSE?

This lava is flowing from Kilauea, Hawai`i's most active volcano. What will happen to the lava as it cools?

SCIENCE VOCABULARY

plate (PLĀT)

A **plate** is a large section of Earth's crust and outer mantle that slowly moves. (p. 195)

Earth's crust is broken into about 20 huge plates.

fault (FAWLT)

A **fault** is a crack in Earth's crust where slabs of rock can slip past, move away from, or push against each other. (p. 196)

Earthquakes usually occur near a fault.

earthquake (URTH-kwāk)

An **earthquake** is the shaking of the ground usually caused by the movement of Earth's crust or by a volcano. (p. 197)

Movement of Earth's crust during an earthquake can damage buildings.

my Science Vocabulary

earthquake (URTH-kwāk)	**magma** (MAG-mu)
fault (FAWLT)	**plate** (PLĀT)
lava (LAH-vu)	**volcano** (vol-KĀ-nō)

magma (MAG-mu)

Magma is melted rock beneath Earth's surface. (p. 200)

Magma reaches the surface through vents in a volcano.

Hawaii

Magma chamber

Lo'ihi Seamount

lava (LAH-vu)

Lava is melted rock that flows from a volcano onto Earth's surface. (p. 200)

Lava flows down the side of this active volcano.

volcano (vŏl-KĀ-nō)

A **volcano** is a mountain or hill on Earth's crust through which lava, gas, and ashes can erupt. (p. 200)

Ash, lava, and gas spew through the air as this volcano erupts.

Earth's Structure

Have you ever bitten into a warm, juicy peach? Your teeth pierce the skin, scrape through the juicy pulp, and stop short of the hard pit inside. Like the peach, Earth is made up of three main layers—the crust, the mantle, and the core. The crust covers Earth like the skin of the peach. It's the hard rocky part we live on. Beneath the crust lies the mantle. The top part of the mantle is solid rock, like the crust. The rest of the mantle is partly melted rock. Next comes the core. The outer core is liquid. The inner core is solid.

These mountains in Asia called the Himalaya, form as two plates push against each other.

EARTH'S LAYERS

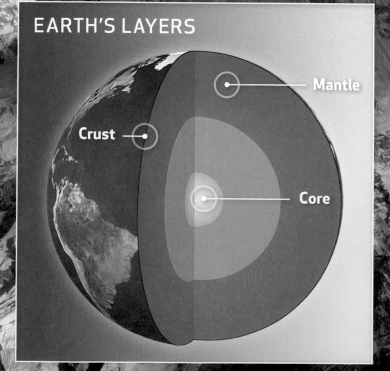

Crust

Mantle

Core

Earth's crust and upper mantle form a solid rocky shell. This shell is broken into about 20 huge rocky slabs called **plates** . Plates can slide past each other, move away from each other, and push against each other. Most of the time this movement occurs slowly—just a few centimeters each year. Over time, this movement reshapes the land and ocean floors. For example, mountains form when plates push against each other.

In Iceland, a crack has formed because two plates are slowly moving away from each other.

Two plates slide past each other near the San Andreas fault in California.

Before You Move On

1. What are Earth's three main layers?
2. Describe the three directions in which plates can move.
3. Infer How is a hard boiled egg like the layers of Earth?

Earthquakes

Look at the line that forms across the land in the photo. Notice how the land on one side of this line is raised. What might have happened to make the land look like this?

The slow movement of Earth's plates can push and pull on the rock in the crust. These forces can crack the rock, making faults . A fault is a crack in the crust where slabs of rock can sometimes slip past each other. The largest faults are at the boundaries of the plates.

When a very large earthquake happened in Southern California in 1992, Earth's crust experienced a break along the fault.

Sometimes part of one plate gets caught on the rough edge of another. The huge slabs of rock can push against each other but don't move. Pressure builds. Suddenly, the plates break free. The release of pressure causes a jolt. Shock waves move out in all directions making the ground shake. It's an earthquake !

Science in a Snap! Pressure Buildup

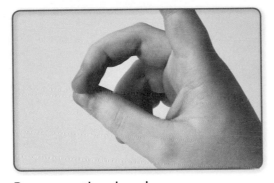

Press your thumb and middle finger together.

Keep pressing as you try to slide your thumb and finger past each other.

What happened as you kept pressing?
How is this similar to what happens during an earthquake?

Thousands of earthquakes happen every day. So why don't we hear about them all? Most are too weak to feel. They can be detected only by instruments called *seismographs*.

Other earthquakes can raise and lower the land. They can even change the course of rivers! Strong earthquakes change more than the land. They cause damage to buildings, bridges, roads, and other structures. Buildings can fall, roads buckle, bridges collapse, and railroads twist.

An earthquake caused major structural damage to buildings in Indonesia.

MOMENT MAGNITUDE SCALE

Each time the magnitude increases by 1 it means the earthquake releases 32 times more energy.

MAGNITUDE					
0–3.9 Minor	4.0–4.9 Light	5.0–5.9 Moderate	6.0–6.9 Strong	7.0–7.9 Major	8.0–above Great
EFFECTS NEAR CENTER OF EARTHQUAKE					
rarely noticed	slight damage	some buildings damaged	major damage to buildings	well-built buildings damaged	major to total destruction

Scientists measure the strength of earthquakes by using the moment magnitude scale. This scale measures the amount of energy that comes from an earthquake. Earthquakes with a low magnitude release little energy and do little damage. An earthquake that measures 1.0 can't be felt by anyone but can be recorded by seismographs. You could feel an earthquake with a magnitude of 4.0. One with a magnitude of 6.0 could damage buildings and roads, and one with a magnitude of 8.0 could be a major disaster.

TECHTREK
myNGconnect.com

Enrichment Activities

These engineers test different materials to make buildings safer during earthquakes.

An earthquake with a magnitude of 7.9 caused this bridge to collapse in southwest China.

Before You Move On

1. In geology, what is a fault?
2. How does an earthquake occur?
3. **Draw Conclusions** How might an earthquake that measures 6.5 be more damaging to people than an earthquake that measures 7.5?

Volcanoes

Sometimes **magma**—melted rock in the mantle—flows up to the surface through openings in Earth's crust. When the hot magma spills onto the surface, it becomes **lava**. The lava hardens into solid rock. This rock can build up over time and forms a mountain—a **volcano**.

Most volcanoes form where plates collide and one plate sinks beneath the other. The sinking plate melts in the mantle and becomes magma. The magma rises through weak spots in the crust. When the magma comes out onto the surface, the volcano erupts. An eruption includes magma, gases, ashes, and chunks of hot rocks.

 Eldfell, or "Fire Mountain" is a volcano in Iceland. It first erupted in January of 1973. In five months Eldfell grew to 215 meters (705 feet) high.

Volcanoes take on different shapes depending on how they formed. Three types of volcanoes are cinder cone, shield, and composite volcanoes. Compare and contrast these kinds of volcanoes.

VOLCANOES

TECHTREK myNGconnect.com

Digital Library

PARÍCUTIN, MEXICO

CINDER CONE VOLCANO

A cinder cone is a small volcano with steep sides. Eruptions are mostly ashes and larger pieces of rock. This material piles up around the opening, or vent, and forms a cone shape.

KILAUEA VOLCANO, HAWAI'I

SHIELD VOLCANO

A shield volcano is a wide volcano with gently sloping sides. Eruptions are mostly liquid lava that spurts in the air or flows out gently.

MOUNT FUJI, JAPAN

COMPOSITE VOLCANO

A composite volcano is cone-shaped, but is much taller than cinder cones. Composite volcanoes are made of layers of lava sandwiched between layers of ashes and rock pieces. Most of the violent eruptions you hear about are from composite volcanoes.

Volcanic eruptions can be very dangerous. Hot ash and lava can cover the land, destroying buildings and burying towns. Volcanic ash can clog lakes and rivers, damage machines, cause breathing problems, and destroy crops. Many eruptions include globs of molten rock that cool in mid air. They become hot rocks called volcanic bombs, which can crash through buildings and cars.

A volcanic bomb in Puy-de-Dôme France.

A giant cloud of ash fills the sky as Mount Pinatubo erupts on the island of Luzon in 1991.

Volcanic eruptions can have some surprising effects on Earth's surface. For example, ash from a volcano can eventually become good soil for growing crops. Lush vineyards and orchards cover the lower slopes of Mount Vesuvius, which destroyed Pompeii and several other towns in A.D. 79. Volcanoes can also create new landforms. The Hawaiian Islands formed as volcanoes built up from the floor of the Pacific Ocean.

This undersea volcano erupted off the coast of Tonga in March 2009. This eruption built a new island in less than a week.

Before You Move On

1. Describe three types of volcanoes.
2. What are some dangers of a volcanic eruption?
3. **Explain** How does a volcano form?

Landslides

Have you ever seen a road sign that says *Caution: Falling Rock?* Rock on a ledge or steep hill may break loose and fall. Sometimes it's just a rock or two. But at other times, it's a landslide.

A landslide is a rapid movement of rock, soil, and other material down a hill or mountain. If enough material falls, the shape of the hill or mountain can be changed a lot in just a few minutes. Landslides are often caused by volcanoes. Volcanic eruptions can set off landslides as their powerful explosions loosen rock from the side of a mountain.

This photo from space shows the volcano Mount St. Helens after it erupted on May 18, 1980. The gray area shows that the side of the volcano exploded and slid downhill during the eruption.

Earthquakes can also set off landslides. The energy released during an earthquake can break free large amounts of rock and soil. Weathering, volcanoes, and earthquakes are triggers. But it is the force of gravity that actually pulls the material downhill.

On October 17, 1989, the Loma Prieta earthquake caused this landslide near San Francisco, California.

Before You Move On

1. What is a landslide?
2. What makes material in any landslide move downhill?
3. **Infer** How can an earthquake trigger a landslide?

NATIONAL GEOGRAPHIC

LIVING ON THE EDGE
IN JAPAN

Japan has more than its share of earthquakes and volcanoes. Earthquakes and volcanoes can't be predicted easily. So, people in Japan learn all they can about them and often practice what to do when they happen.

Think about the things in your home. There are lots of things that are heavy or could do damage if they fell on someone. In Japan, people are told to keep these things put away where they can't hurt anyone. Families also keep survival kits handy—bottled water, food, flashlights, and first-aid kits. They know how to turn off gas and electricity after a disaster.

Check dams like these are built to hold back or slow mudflows that can occur as a result of volcanic eruptions. These dams give people more time to seek shelter.

Students in schools prepare for natural disasters. Earthquake drills are held at many schools once a month. Students learn how to get under their desks during an earthquake. If they're playing outside, students learn how to gather away from the school building. Sometimes, students even get to practice in a special room that feels just like the Earth is shaking.

Scientists experiment to find new ways to help. They keep accurate records on earthquake activity. They keep careful watch on any ground movement, well water levels, and other signs that an earthquake or eruption might be coming. They are hoping to find patterns of small changes that can help them predict a larger earthquake. Scientists are also hoping to use these same strategies to figure out when an active volcano may erupt next.

These students wear protective headgear during an earthquake drill.

Students learn first-aid to use during a possible emergency.

Conclusion

Earth's surface is always changing. Mostly these changes happen slowly over a long period of time. However, change can also happen quickly. Earthquakes can shake the ground, making buildings fall. Volcanic eruptions can create new land or destroy existing land. Finally, landslides can cause hillsides to move downhill.

Big Idea Earthquakes, volcanoes, and landslides can change Earth's surface quickly.

QUICK CHANGES TO THE EARTH'S SURFACE CAN BE CAUSED BY:

Earthquakes

Volcanic Eruptions

Landslides

Vocabulary Review

Match the following terms with the correct definition.

A. earthquake

B. fault

C. lava

D. magma

E. plate

F. volcano

1. An opening in Earth's crust through which lava, gases, and ash erupts

2. A crack in Earth's crust where slabs of rock can slip past each other

3. Melted rock beneath Earth's surface

4. A shaking of the ground usually caused by the sudden movement of Earth's crust or by a volcano

5. Melted rock that comes to Earth's surface through a volcanic eruption

6. A large section of Earth's crust and outer mantle that slowly moves

Big Idea Review

1. **Identify** In which of Earth's layers do earthquakes and volcanoes occur?

2. **Describe** Describe how earthquakes can change Earth's surface.

3. **Explain** What role does gravity play in a landslide?

4. **Compare and Contrast** What is the difference between lava and magma?

5. **Analyze** How can volcanoes destroy land and create land?

6. **Draw Conclusions** An earthquake measuring 6.0 on the moment magnitude scale occurs near a city. What buildings will be most affected? Which will be least affected?

Write About Earth's Surface

Infer What is happening in this diagram? How is this earthquake changing Earth's surface? How will it also affect the cities? Will one be affected more than the others?

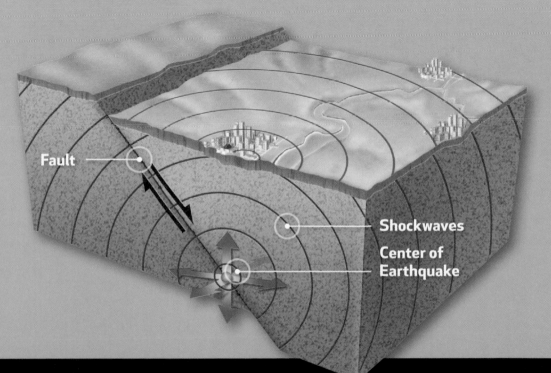

Fault

Shockwaves

Center of Earthquake

EARTH SCIENCE EXPERT: VOLCANOLOGIST

Are you interested in volcanoes? How about being a volcanologist?

Volcanoes can be beautiful and exciting. They are Earth's most dangerous type of landform! Maybe you would like to climb one or study what's inside. Or you might want to help save lives by learning to predict eruptions. If you are someone like this, then you might want to be a volcanologist like Tamsin Mather.

What does a volcanologist do?

I study volcanoes and their effects on our planet. Some of my time is spent visiting volcanoes to take measurements and collect samples. I take the samples to a lab for analysis. I also work in a university. I teach students about volcanism and Earth science.

Tamsin Mather studies the Villarrica Volcano in Chile, South America.

TECHTREK
myNGconnect.com

Digital Library

Tamsin and a fellow scientist set up equipment to study a volcano in Nicaragua, Central America.

What is a typical day like?

This depends on whether I am in the field or not. Field days usually start early. First, I get the equipment ready. Then, I hike to the site to collect samples. Once back from the field I unpack my samples and write notes about my day. On days when I'm not in the field, I teach classes and meet with students. I also spend time in the lab.

Do you see a strong connection between what you do and Earth Science?

Of course! Volcanoes are central to the Earth sciences. Volcanoes have played a key part in the history of our planet. They continue to shape our planet today. Major eruptions show the importance of Earth science to people's safety.

Tamsin wears a mask to protect her from gases coming from the volcano, Mount Etna, in Italy.

BECOME AN EXPERT

The Hawaiian Islands:
Formed by Volcanoes

Most **volcanoes** and **earthquakes** happen along **faults** . But that's not always the case. Take Hawai'i, for instance. Hawai'i is the home to six major volcanoes, including the biggest and most active volcano on Earth. In fact, the Hawaiian Islands are the tops of large volcanoes that built up from the ocean floor.

Mt. Everest is the highest mountain on Earth, but Kilauea is taller if measured from its base on the ocean floor.

volcanoes

A **volcano** is a mountain or hill on Earth's crust through which lava, gas, and ashes can erupt.

earthquakes

An **earthquake** is the shaking of the ground usually caused by the movement of Earth's crust or by a volcano.

TECHTREK
myNGconnect.com

Student
eEdition

Digital
Library

The Hawaiian Islands are a mix of active and extinct volcanoes. Active volcanoes are either erupting or show signs that they may erupt in the near future. Extinct volcanoes no longer erupt. The most active volcano in the world, Mauna Loa, is found on the biggest of the Hawaiian Islands, called Hawai`i. Kilauea, is another volcano on Hawai`i. Mauna Loa and Kilauea are shield volcanoes, which form when liquid **lava** flows and hardens into thin layers of rock.

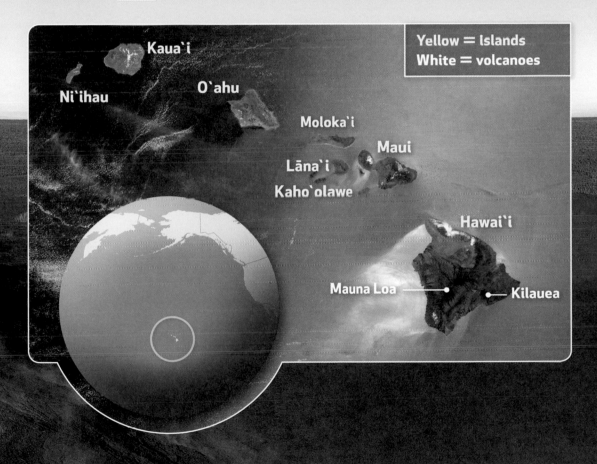

Yellow = Islands
White = volcanoes

Kaua`i

Ni`ihau

O`ahu

Moloka`i

Maui

Lāna`i

Kaho`olawe

Hawai`i

Mauna Loa

Kilauea

fault

A **fault** is a crack in Earth's crust where slabs of rock can slip past, move away from, or push against each other.

lava

Lava is melted rock that comes to Earth's surface by a volcanic eruption or lava flow.

Hot Spots

The Hawaiian Islands are far from the edge of a **plate** . In fact, they're in the middle of a plate that covers most of the Pacific Ocean. Then how did these volcanic islands form? They formed over a hot spot. Sometimes **magma** in the mantle rises through a weak spot in the crust. It's called a hot spot. The diagrams below show how a hot spot forms, and continues to form, the Hawaiian Islands.

HOT SPOTS FORM ISLAND CHAINS

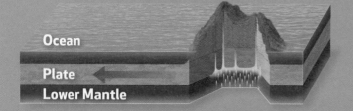

STEP 1 Magma rises through the plate and forms a volcano.

STEP 2 As the plate moves, the volcano moves with it. The hot spot stays in the same place.

STEP 3 The first volcano stops erupting and becomes an extinct volcano. A second volcano forms over the hot spot.

STEP 4 The plate continues to move. A third volcano forms over the hot spot.

plate

A **plate** is a large section of Earth's crust and outer mantle that slowly moves.

magma

Magma is melted rock beneath Earth's surface.

The chain of Hawaiian Islands formed in this way. Now look at the diagram below. It shows the Hawaiian islands and the direction the plate is moving over the hot spot. Which island do you think formed first? Ni`ihau did. One of Hawaii's youngest volcanoes is Lō'ihi seamount. A seamount is a mountain or volcano that is completely under the ocean. Find Lō'ihi below. Predict what will eventually happen to Lō'ihi.

PREDICTING HOT SPOT ACTIVITY

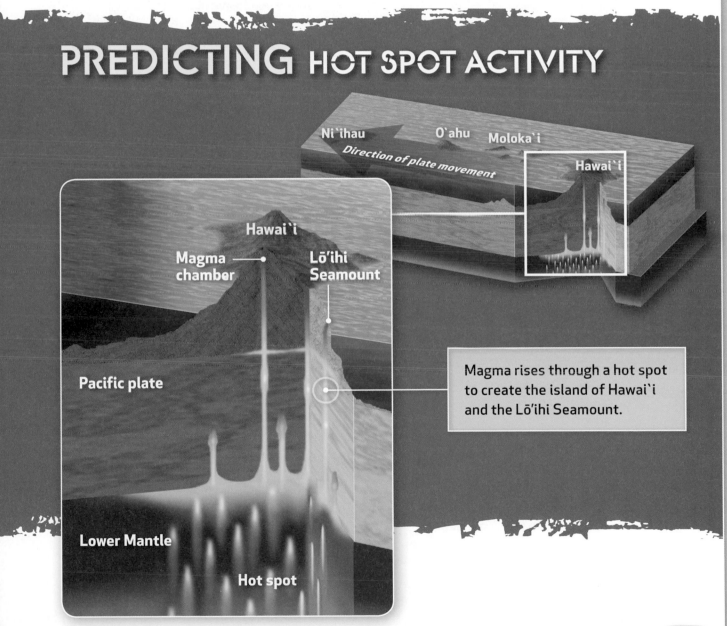

Ni`ihau    O`ahu   Moloka`i

Direction of plate movement

Hawai`i

Hawai`i

Magma chamber

Lō'ihi Seamount

Pacific plate

Magma rises through a hot spot to create the island of Hawai`i and the Lō'ihi Seamount.

Lower Mantle

Hot spot

Visiting Hawai`i

If you visited the Hawaiian Islands, you would see things there you can't see anywhere else in the United States. You could learn about the volcanic features found on these islands. One of these features are the black sand beaches. The black sand is actually bits of lava rock. The waves have battered and broken the rock into grains of sand over thousands of years.

Black sand is made up of tiny shards of lava rock, which is formed by a fast change—explosive cooling of lava in cold ocean water.

This is the Punalu`u black sand beach on Hawai`i.

What happens when lava meets water? The water instantly boils! Some volcanoes on Hawai`i have side vents that are close to water. Steam rises and lava splatters as the hot molten rock cools quickly in the water.

Lava tubes also fascinate visitors. These cave-like structures form during lava flows that last a long time. When the lava flows along the same path, it starts to solidify and build up into walls. A roof might form as well. Lava can then travel along these tubes for long distances. Between 1969 and 1974, lava flows from Mount Kilauea traveled through a system of lava tubes for more than 11 kilometers (7 miles)!

The Thurston lava tube is part of the system of lava tubes from Mount Kilauea, in the Hawaiian Islands.

TECHTREK
myNGconnect.com

Digital Library

Lava flows into the Pacific Ocean at Hawai`i Volcanoes National Park.

The Emperor Seamounts

Northwest of the oldest Hawaiian island, Kaua`i, scientists have found a line of underwater mountains. They are called the Emperor Seamounts. Scientists think these mountains were caused by the same hot spot that formed Hawai`i. The hot spot started forming these volcanoes about 70 million to 42 million years ago. As they moved away from their original location, they started to erode, or wear away, until they were underwater.

FORMATION OF SEAMOUNTS

Volcano

Coral Reef

Lagoon

Seamount

Over time, island volcanoes erode until they are underwater.

Notice that the Emperor Seamounts run north-south. The Hawaiian Chain runs northwest-southeast. Why is there a bend in the direction of these features? Scientists think the plate started to move differently at one point in the past.

New volcanoes will continue to form as the islands of Hawai`i move away from the hot spot. The current residents of Hawai`i don't need to worry about a new address, though. These kinds of changes will take another few million years!

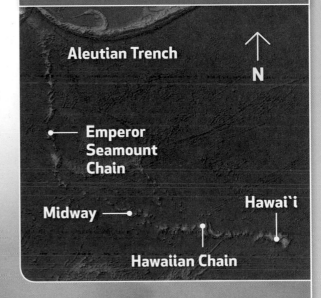

This map shows the floor of the Pacific Ocean. Notice that the Emperor Seamount forms a straight line and points almost straight north.

Aleutian Trench

N

Emperor Seamount Chain

Midway

Hawai`i

Hawaiian Chain

CHAPTER 5
SHARE AND COMPARE

Turn and Talk How did the Hawaiian Islands form? Form a complete answer to this question together with a partner.

Read Select two pages in this section. Practice reading the pages. Then read them aloud to a partner. Talk about why the pages are interesting.

Write Write a conclusion that tells the important ideas you learned about hot spots and the Hawaiian Islands. State what you think is the Big Idea of this section. Share what you wrote with a classmate. Compare your conclusions. Did your classmate recall that the island of Hawai`i will eventually move away from the hot spot?

Draw Form groups of four. Have each person draw a picture of a different volcanic feature on Hawai`i. Add labels to your drawings. Put the drawings together to create a display of volcanic features on Hawai`i.

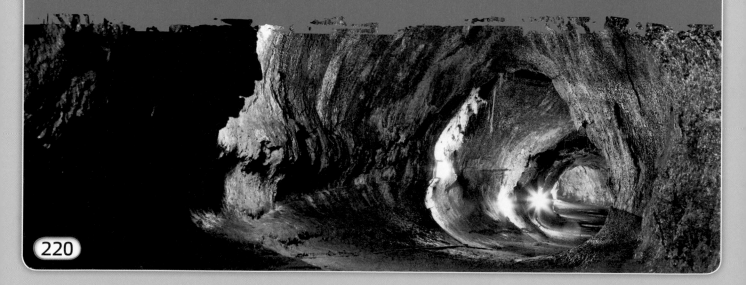

CHAPTER
6

WHAT CAN WE OBSERVE ABOUT WEATHER?

The look of the sky can tell you a lot about the weather. It can also help you predict the weather to come. Find the huge, dark cloud off in the distance. What can you observe about this cloud? When you see a cloud like this, you know stormy weather is on the way.

TECHTREK
myNGconnect.com

Storm clouds like this one can reach 20 kilometers (12 miles) high. Fierce winds often come with storm clouds.

222
223

After reading Chapter 6, you will be able to:

- Identify and describe weather changes and patterns by measurable quantities.
  PARTS OF WEATHER, CLOUDS, AIR MASSES AND WEATHER FRONTS

- Explain that air surrounds us and takes up space. **AIR AND THE ATMOSPHERE**

- Identify and describe temperature, precipitation, air pressure, wind and humidity as measurable quantities of weather. **PARTS OF WEATHER**

- Describe how water exists in the air and how clouds form. **CLOUDS**

- Identify and describe how air masses and fronts affect weather.
  AIR MASSES AND WEATHER FRONTS

- Identify and describe weather patterns that affect the United States.
  AIR MASSES AND WEATHER FRONTS

- Science in a Snap! Explain that air surrounds us and takes up space. **AIR AND THE ATMOSPHERE**

CHAPTER 6

WHAT CAN WE OBSERVE ABOUT WEAT

The look of the sky can tell you a lot about the weather. It can also help you predict the weather to come. Find the huge, dark cloud off in the distance. What can you observe about this cloud? When you see a cloud like this, you know stormy weather is on the way.

TECHTREK
myNGconnect.com

Student
eEdition

Vocabulary
Games

Digital
Library

Enrichment
Activities

HER?

Storm clouds like this one can reach 20 kilometers (12 miles) high. Fierce winds often come with storm clouds.

SCIENCE VOCABULARY

atmosphere (AT-mus-fear)

Atmosphere is the layer of gases around Earth. (p. 226)

Earth's atmosphere is where weather occurs.

humidity (hyū-MID-i-tē)

Humidity is the water vapor in the air. (p. 237)

The amount of humidity in the air affects the weather.

evaporation (ē-vap-u-RĀ-shun)

Evaporation is when matter changes from a liquid to a gas. (p. 238)

Evaporation of water from the ocean makes the air feel humid.

my
Science Vocabulary

air mass
(air mas)

atmosphere
(AT-mus-fear)

condensation
(kon-den-SĀ-shun)

evaporation
(ē-vap-u-RĀ-shun)

front
(FRUNT)

humidity
(hyū-MID-i-tē)

TECHTREK
myNGconnect.com

Vocabulary
Games

condensation
(kon-den-SĀ-shun)

Condensation is when matter changes from a gas to a liquid. (p. 238)

Condensation occurred when the moist air cooled as it touch the leaf.

air mass (air mas)

An **air mass** is a large body of air that takes on the temperature and humidity of the land over which it forms. (p. 246)

A warm, moist air mass brings hot and humid weather.

front (FRUNT)

A **front** is the place where two different air masses meet. (p. 248)

Thunderstorms often form at cold fronts.

Air and the Atmosphere

Look closely at Earth's atmosphere in this photo. The atmosphere is the layer of gases around Earth. Most living things need the gases in the atmosphere to live. The atmosphere acts like a blanket holding heat near Earth's surface. All weather occurs in the atmosphere, too. Although you can see the atmosphere in this photo, you don't usually see the air around you. So how do you know it's there? You feel it around you as wind. You can also gather evidence to show that it takes up space.

Earth's atmosphere forms a layer of gases about 600 kilometers (370 miles) thick around the planet.

Crumple a piece of paper. Tape the paper to the inside bottom of a cup.

Turn the cup upside down and push it straight down into a bowl of water. Keep the cup straight as you pull it out of the water. Observe the paper.

What do your observations about the paper tell you about air?

The atmosphere is made up of five layers that have different properties. More than three quarters of the atmosphere's air is in the layer closest to Earth's surface. The atmosphere is made of several gases. About 78 percent is nitrogen and about 21 percent is oxygen. The remaining 1 percent contains several other gases.

Water vapor makes up only a small percentage of the atmosphere. But it's a very important gas. It is one of the gases that helps Earth hold in heat. It also forms clouds, rain, and snow.

This picture of the sun setting was taken from the International Space Station. The light in the lower layers of the atmosphere looks red and orange. The light in the upper layers looks blue.

LAYERS OF **EARTH'S ATMOSPHERE**

In this outermost layer, the air particles are very far apart. The atmosphere just fades away to space.

ABOVE 550 KM (342 MI)

Air in this layer has a higher temperature than a furnace that's used to melt steel!

ABOVE 80 KM (50 MI)

The coldest temperatures occur in this part of the atmosphere.

50 TO 80 KM (31 TO 50 MI)

This layer includes a form of oxygen called ozone that absorbs harmful radiation from the sun.

12 TO 50 KM (7 TO 31 MI)

Most of the air is in this first layer. Weather, including clouds, storms, wind, rain, and snow, occurs here.

0 TO 12 KM (0 TO 7 MI)

EARTH

Before You Move On

1. What is the atmosphere?
2. Why is water vapor an important part of the atmosphere?
3. **Cause and Effect** What would Earth be like without an atmosphere?

Parts of Weather

Have you ever left your house on a sunny day without an umbrella—only to get caught in the rain? If so, you know that the weather can change fast. Weather is the way the atmosphere is at a certain place at a certain time. Weather can change from day to day, and even hour to hour. In one day, weather can go from gray skies and snow to clear skies and bright sun.

What parts of the weather do you think this girl is feeling?

Temperature, precipitation, air pressure, wind, and humidity determine what our weather will be like. And they change all the time. When you observe the weather, you usually observe one or more of these things. Is it hot or cold? Is it windy or rainy? Scientists use the tools in the chart to observe the weather.

MEASURING THE WEATHER

TEMPERATURE

A thermometer measures temperature.

PRECIPITATION

A rain gauge measures rainfall.

AIR PRESSURE

A barometer measures air pressure.

WIND

A wind vane shows wind direction.

WIND

An anemometer measures wind speed.

HUMIDITY

A hygrometer measures humidity.

Temperature When you think about weather, you probably think about how hot or cold it is outside. Temperature is the measurement of how hot or cold something is, such as the outside air. Temperature can change all the time. It is usually warmer during the day than during the night. The temperature can also change depending on the time of year. Winters are usually colder than summers. Have you ever climbed a mountain? Temperature can also change as you climb higher. High places are usually cooler than low places.

Weather can be snowy when the temperature is low.

Precipitation Precipitation is water or ice that falls from clouds. There are five main types of precipitation—rain, freezing rain, sleet, snow, and hail. Each type forms under different conditions. Most precipitation starts as ice crystals in the coldest parts of clouds. Even rain often starts in the clouds as snow. The type of precipitation that reaches the ground depends on the temperature of the air between the cloud and the ground.

TYPES OF **PRECIPITATION**

TECHTREK
myNGconnect.com

Digital Library

RAIN Ice crystals in clouds become heavy enough to fall. They melt as they fall through a layer of warmer air between the cloud and the ground. Precipitation reaches the ground as rain.

FREEZING RAIN Ice crystals in clouds melt into water drops as they fall through a layer of warmer air. Then they reach a layer of freezing air (below 0°C) near the ground. The cold water drops freeze onto cold objects such as trees and roads as they touch them.

SLEET Ice crystals from clouds melt as they drop through a layer of warm air high off the ground. They then fall through a layer of colder air closer to the ground. This freezes the water drops into particles of ice. Sleet then hits the ground.

SNOW Ice crystals from clouds fall through air that is below freezing on the way to the ground. Precipitation reaches the ground as snowflakes.

HAIL Lumps of ice form in some clouds. They add layers of ice as moving air in the clouds bounces them up and down through very cold water. The hailstone grows until it is heavy enough to fall from the cloud.

Air Pressure Suppose you put a book on your head. You would feel it pressing down. Did you know that air also presses down on you and everything else on Earth? Particles of air are tiny. You don't feel them as you walk around. However, the layer of air around Earth goes up for many kilometers. When you put all of those particles of air together, they have a lot of weight. The weight of the air is called air pressure.

When air pressure rises or goes from low to high, then fair weather is coming.

The amount of air pressure is different from place to place. In some places, air pressure is high. This means the air is heavier. The air particles are close together. In other places, air pressure is low. There are fewer air particles, and they are more spread out. The air is lighter. Changes in air pressure can tell you that the weather is about to change.

When the air pressure changes from high to low, rainy weather may be on the way.

Wind You know that air is made up of tiny particles. These air particles are constantly moving. You don't usually feel this movement. But when lots of air particles move together from one place to another, you do feel it. This moving air is called wind. What causes air to move from one place to another? Differences in air pressure make air move. Air moves from areas of higher pressure to areas of lower pressure.

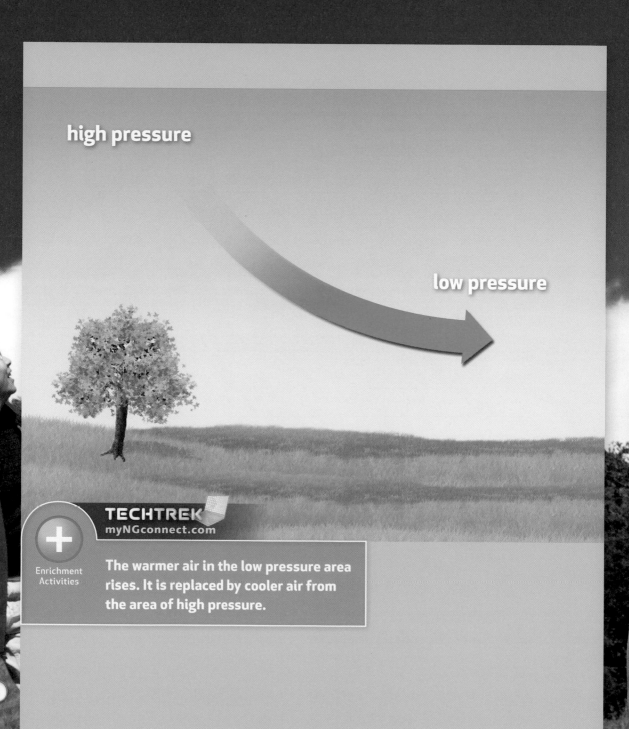

high pressure

low pressure

TECHTREK
myNGconnect.com

Enrichment
Activities

The warmer air in the low pressure area rises. It is replaced by cooler air from the area of high pressure.

Humidity On a very hot day, your skin may feel sticky. That's because warm air holds a lot of water vapor. The water vapor in the air is called humidity . Cold air can't hold as much water vapor as warm air can. So, warm air can be more humid than cold air. Knowing the amount of humidity helps to predict weather. For example, high humidity can cause clouds to form. Clouds can produce precipitation.

Before You Move On

1. What are the five major things that affect weather?
2. What happens in the atmosphere to produce freezing rain?
3. **Analyze** There's a snowstorm outside. Tell what is happening with the five parts of weather.

Clouds

Clouds are important in the water cycle. This cycle moves water from Earth's surface to the atmosphere and back again. As the sun's energy causes water on Earth's surface to get warmer, evaporation occurs. Evaporation is when matter changes from a liquid to a gas. When water evaporates, it forms a gas called water vapor. Water vapor rises. As it rises, it cools. The cooling of water vapor causes condensation . Condensation is when matter changes from a gas to a liquid.

CLOUD SHAPES

cumulus clouds

stratus clouds

When water vapor condenses in the air, it forms very tiny drops of water. These tiny drops make up many clouds. Cloud droplets are so small that they float on air. In fact, about one million tiny cloud droplets are needed to make just one raindrop. There are many types of clouds. But there are just three basic shapes—cumulus, stratus, and cirrus. Each type of cloud forms in different conditions.

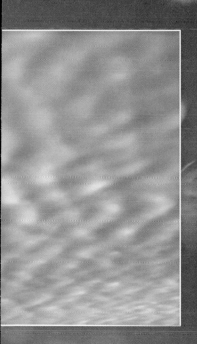

cirrus clouds

Water vapor cools when it touches the leaf.
When it cools, it condenses and forms dew.

Cumulus Clouds Cumulus clouds are white, puffy clouds that look like cotton balls. On clear, warm days, they start to build in late morning as the sun heats the ground. The clouds form as water condenses at the tops of warm, moist columns of air that rise from the ground. Cumulus clouds are fair weather clouds. But keep an eye on them on warm afternoons. They can turn into storm clouds.

Cumulus clouds form where warm, moist air rises.

On warm, sunny afternoons, puffy cumulus clouds can thicken. They can grow taller and become cumulonimbus clouds. These are rain clouds that grow from cumulus clouds and form a flat top where they meet the next layer of the atmosphere. When you see these clouds, look for short periods of stormy weather. As they approach, strong winds begin to blow. Heavy rain, lightning, and thunder follow. In fact, cumulonimbus clouds are sometimes called thunderheads.

Cumulonimbus clouds can be very tall—sometimes 12 kilometers (7 miles) or more.

Stratus Clouds Some days are gloomy outside, and the sky is covered with gray clouds. These clouds are stratus clouds. Stratus clouds are low and flat. They form a layer of gray across the sky. Often, stratus clouds just bring cloudy weather. But sometimes, they cause light drizzle. Fog is also a stratus cloud. It forms close to the ground.

The stratus clouds in this picture are so close to the ground that they are covering the top of these rock formations called buttes.

When stratus clouds thicken, they can become nimbostratus clouds. These are the dark gray clouds that form precipitation. Sometimes nimbostratus clouds cover the whole sky. Recall that cumulonimbus clouds cause short, heavy rain showers. Nimbostratus clouds cause a longer period of lighter rain and gray skies. In fact, nimbostratus clouds are often called rain clouds. In colder weather, they can bring snow.

You can't see the sun or moon through nimbostratus clouds.

Cirrus Clouds These thin, white streaks are cirrus clouds. Look for cirrus clouds in fair weather. Cirrus clouds are the highest clouds. They form high in the first layer of the atmosphere where it is very cold. As a result, cirrus clouds are made of tiny ice crystals. When strong winds high up blow the ice crystals around, cirrus clouds have curved tails.

Cirrus clouds form higher in the atmosphere than either cumulus or stratus clouds.

Cirrus clouds do not produce rain. But when you see them in the sky, a change in the weather is on the way. Gradually, cirrus clouds will give way to lower, thicker clouds. In time, stratus clouds will cover the sky. The weather can then turn rainy or snowy.

Jet planes can create cirrus clouds called contrails. When the planes fly, they release hot gas full of water vapor. The water vapor turns into small ice crystals in the cold air and looks like smoke coming from the tail of the Jet.

Before You Move On

1. What are the three main types of clouds?
2. What is the difference between cumulus and cumulonimbus clouds?
3. **Predict** Suppose you look up in the sky and see stratus clouds. What kind of weather could you expect?

Air Masses and Weather Fronts

When an **air mass** passes over the area where you are, the weather will usually stay the same for two or three days. An air mass is a large body of air that takes on the temperature and humidity of the land over which it forms. An air mass is huge. It can be more than 1,000 kilometers (600 miles) wide. That's wider than the state of Montana!

A cold and dry land air mass was moving over the area where this picture was taken, causing the weather to be cold and dry.

Air masses are not all the same. They can be cold or warm. They can be humid or dry. When an air mass moves over an area, it brings its own weather with it. If a cold, dry air mass came from the North Pole, it would make an area cold and dry. If a warm and humid ocean air mass moved onto land, it would become warm and humid in that area.

The map shows two of the different types of air masses. The blue area shows a cold and dry air mass from the North Pole. The orange areas show warm and humid air masses that formed over the ocean near the Equator.

Cold Fronts Air masses usually stay apart, but sometimes they meet. The border where two air masses meet is called a front. What happens when a cold air mass runs into a warm air mass? The cold air pushes under the warm air, forcing the warm air up. This is a cold front. As a cold front gets closer, you might see a line of tall cumulonimbus clouds in the distance. This is the front line. Gusty winds may start to whip around as the front gets closer. The skies will darken, as the tall storm clouds move overhead.

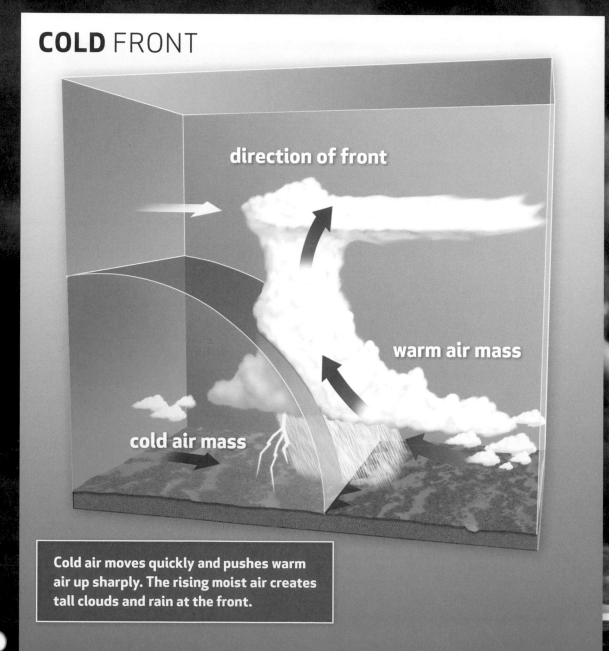

COLD FRONT

direction of front

warm air mass

cold air mass

Cold air moves quickly and pushes warm air up sharply. The rising moist air creates tall clouds and rain at the front.

These tall clouds will bring a heavy downpour. There will also be lightning and thunder. Storms at cold fronts are strong. But they pass quickly, because cold fronts move fast. After the cold front passes, a cooler and drier air mass follows. It brings clear skies and cooler temperatures.

Strong thunderstorms can form at cold fronts.

Warm Fronts What happens when a warm air mass runs into a colder air mass? The warm air glides up over the cold air. This creates a warm front. Like cold fronts, warm fronts produce clouds and rain. But the weather change is more gradual. First, you see high cirrus clouds. As the front slowly moves toward you, the sky becomes more gray. Low stratus clouds replace high cirrus clouds.

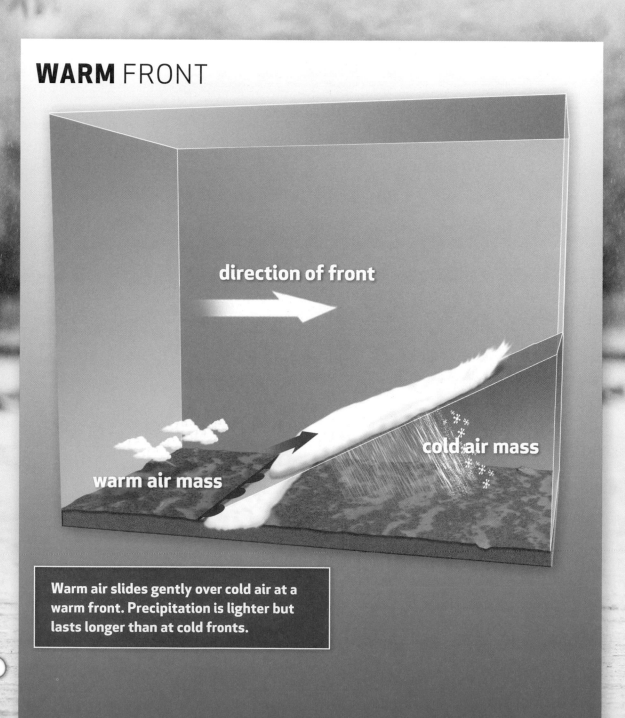

WARM FRONT

direction of front

cold air mass

warm air mass

Warm air slides gently over cold air at a warm front. Precipitation is lighter but lasts longer than at cold fronts.

If the warm air is dry, a warm front means cloudy weather. But moist air can mean light rain. Warm fronts move more slowly than cold fronts. The air also rises more gradually at warm fronts. Clouds increase, sometimes over a couple of days. Then rain or snow falls. After a warm front passes, the temperature rises because the air mass behind a warm front is a warm air mass.

Warm fronts produce rain, but not violent storms.

Weather Patterns Where do air masses come from? They form in regions where the temperature and humidity remain the same most of the time. The air mass takes on the properties of the land or water over which it forms. For example, an air mass that forms over northern Canada will be cold and dry like the land under it. When this air moves to another place, it brings cold and dry weather to the lands it passes over.

This air mass brings cold, dry winter weather to the northeast and midwest.

This air mass brings cool, rainy winters to the northwest coast.

Seattle

This air mass brings precipitation and warm, humid weather to the western part of the United States.

Los Angeles

This air mass causes hot, dry summers in the southwest desert and drought in the central plains.

The map shows some of the important air masses that affect weather in the United States. Continental air masses form over land and bring dry air. Maritime air masses form over water and bring humid air. Polar air masses form over cold, polar regions. They are cold. Tropical air masses form over the warm tropics. They are warm or hot.

This air mass brings cool, rainy winters to the northeast coast.

This air mass brings precipitation and warm, humid weather to the southeastern part of the United States.

• Boston

• Chicago

Houston

• Miami

Before You Move On

1. What is an air mass?
2. What is the difference between a warm front and a cold front?
3. **Draw Conclusions** Find the region where you live on the map on these pages. Where do the air masses come from that most likely affect your weather? How do they affect it?

CLIMBING WEATHER

Even in summer, snow covers the top of Mount Rainier in Washington State. Weather changes fast near the top. A clear day can suddenly become windy, with blowing snow.

Many people climb the mountain each year. They dress in heavy boots and warm clothing. They take ropes, ice axes, and other climbing gear to hike snowy trails and cross glaciers. They also watch the weather—closely.

Climbing Mount Rainier can be risky between September and May because weather can change quickly. Climbers take tents, food, fuel, and sleeping bags to prepare for survival in strong winds and cold.

The first step before a climb is to check the weather on the mountain. A fierce storm near the peak can trap climbers and expose them to days of bitter cold. Experienced climbers know that even clear weather can quickly turn stormy on the mountain. Climbers who spend many days on the trail must check the weather often.

Climbers might look at a weather map like this one. They want clear weather. What can they infer about the weather on the mountain now? It looks clear and cold. Still, a cold front is moving toward the mountain. It could reach Mount Rainier in a day. How might the weather change when the front arrives? Should climbers go now, or wait for a few days?

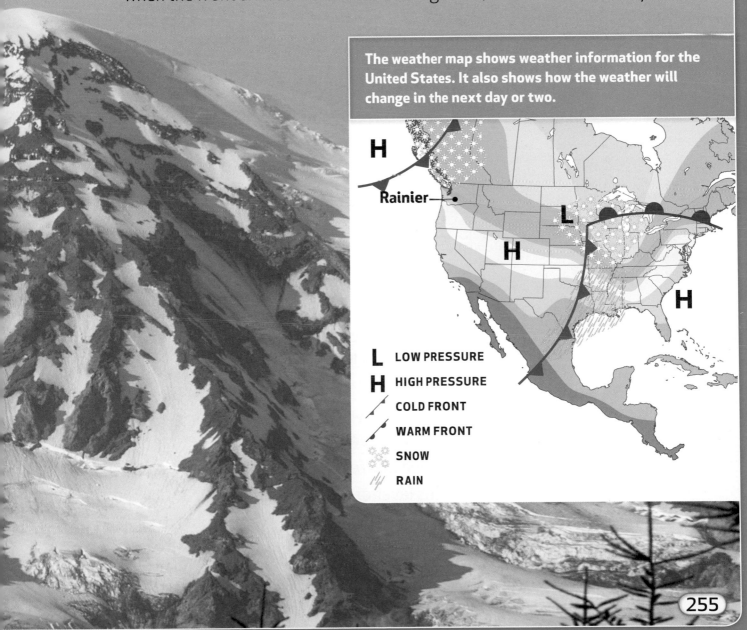

The weather map shows weather information for the United States. It also shows how the weather will change in the next day or two.

Rainier

L LOW PRESSURE
H HIGH PRESSURE
COLD FRONT
WARM FRONT
SNOW
RAIN

Conclusion

The atmosphere is a blanket of gases that surround Earth. Weather occurs in the layer closest to Earth's surface. Temperature, humidity, wind, air pressure, and precipitation are parts of weather. They can be used to predict weather. The parts of weather can also be used to describe properties of air masses and what will happen when air masses meet along fronts.

Big Idea You can observe the atmosphere and use weather instruments to determine what the weather is now and predict how it will change.

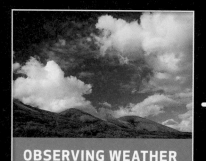

OBSERVING WEATHER

+

USING WEATHER INSTRUMENTS

=

Understanding and predicting weather

Vocabulary Review

Match the following terms with the correct definition.

A. atmosphere

B. humidity

C. evaporation

D. condensation

E. air mass

F. front

1. The water vapor in the air
2. The place where two different air masses meet
3. When matter changes from a gas to a liquid
4. The layer of gases around Earth
5. When matter changes from a liquid to a gas
6. A large body of air that takes on the temperature and humidity of the land over which it forms.

Big Idea Review

1. **Describe** How does the atmosphere change as you go farther away from Earth's surface?

2. **Restate** Use your own words to tell what causes wind.

3. **Contrast** How do the formation of rain and snow differ?

4. **Cause and Effect** You see cirrus clouds in the sky. Should you have your picnic now or wait until tomorrow? Explain your answer.

5. **Apply** Summers in southern Florida are very hot and humid. Where do you think most air masses that move over Florida in the summer come from?

6. **Predict** Suppose it's spring and you look at a weather map. The map shows that a warm front is going to move into the area where you live. What could you predict about the weather in the next few days?

Write About Weather

Draw Conclusions Tell about the parts of weather you observe. Draw conclusions about the temperature, how long it might rain, and what might have caused this weather.

CHAPTER 6

EARTH SCIENCE EXPERT: METEOROLOGIST

What's it like to do a weathercast? Ask Veronica Johnson!

If you live near Washington, D.C., you may know Veronica Johnson. She is a broadcast meteorologist. She does the weather on the four o'clock news each day on TV.

People rely on broadcast meteorologists to find out whether to take a warm coat or an umbrella to school or work. Broadcast meteorologists also warn the public about severe weather, such as hurricanes and tornadoes.

VERONICA JOHNSON'S DAY

- ✓ Look at the sky to see if yesterday's forecast was correct.
- ✓ Observe data on the computer that shows where air masses and fronts are.
- ✓ Check Doppler radar for nearby rain
- ✓ Observe satellite data to track movement of storms
- ✓ Write the forecast to say on the air
- ✓ Make graphics that help tell the weather story

  "Everything is checked off. Now it's show time"!

TECHTREK
myNGconnect.com

Student
eEdition

Digital
Library

It's no surprise that Johnson became a meteorologist. "I was always looking up at the sky and wondering why clouds looked so different from one day to the next," she says. But being on TV still amazes her. Johnson remembers being a good student—but a shy one. She never wanted to talk in class. She never dreamed she would talk to millions of people about the weather on TV!

Johnson says that study of the weather is important because it affects the life and health of everyone. Think you would be interested in broadcast meteorology? Learn as much science and math as you can in school. Then you will need a college degree in meteorology or atmospheric sciences. You must also know how to write and speak well to make a clear presentation on TV and radio.

TECHTREK
myNGconnect.com

Digital
Library

Johnson reviews satellite images like this one to see where severe weather is located and figure out where it is headed next.

Veronica Johnson stands in front of a blank screen in the studio. The weather maps that you see on TV are made by computers—they are not really there. Johnson watches a monitor to make sure she is pointing to the right areas.

BECOME AN EXPERT

Extreme Weather

In some places on Earth, landforms and the **atmosphere** create unusual weather. As a result, these places experience some of Earth's weather extremes.

The Driest The driest place on Earth is the Atacama Desert in Chile, South America. The average rainfall there is less than one millimeter per year. That's about the thickness of a dollar bill. Why is the Atacama so dry? Like most large deserts, the Atacama is under a broad area of sinking dry air. Clouds can't form in sinking air that holds little water vapor. So there is little chance for precipitation.

According to records, some places in the Atacama Desert in Chile have never had rainfall.

atmosphere
The **atmosphere** is the layer of gases around Earth.

TECHTREK
myNGconnect.com

Student
eEdition

Digital
Library

The Hottest   What's the hottest temperature ever recorded on Earth's surface? It's 59°C (138°F). That's about as hot as a bowl of soup or a cup of hot chocolate. It was recorded in Libya in 1922. The spot is Al' Aziziyah, in the Sahara. It's a hot spot because it gets very direct sunlight almost all year. Its elevation is low. The air particles are packed close together and the air holds in heat. No cooling winds blow and no bodies of water to cool the air are nearby.

The hottest temperature ever recorded in the United States was slightly lower. It was 57°C (134°F). That temperature was recorded in California's Death Valley.

The world's highest temperature was recorded in the world's largest hot desert!

The Coldest The coldest temperature ever recorded was –89°C (–128°F). The record was set in 1983 at the Vostok research station in Antarctica. The fact that Antarctica is very cold isn't surprising. Because it's at the South Pole, it gets little to no sunlight half the year. Ice and snow cover most of the continent. They reflect much of the sun's energy that falls on Antarctica. The air over Antarctica is also cold. All of these conditions work together to create a very cold environment.

In which state do you think the coldest temperature in the United States was recorded? The thermometer dropped to –62°C (–80°F) at Prospect Creek, Alaska, in 1971.

This is Antarctica, where the coldest temperature was ever recorded. It doesn't snow much each year in Antarctica. But the snow that falls builds up year after year. The temperatures are too low to melt it.

The Snowiest In the United States, the record for the snowiest winter season goes to Mount Baker in Washington State. In the winter of 1998–99, the area set a world record for snowfall with 29 meters (32 yards). That's enough snow to cover a nine-story building! Why so much snow on Mount Baker? The mountain is near the Pacific coast. Clouds full of moisture blow in from the Pacific. They must rise to get over the mountains. As they do, the temperature inside the clouds drops. **Condensation** occurs, and the clouds drop their moisture as snow.

Because of cold temperatures, precipitation falls on Mount Baker in winter as snow instead of rain.

WHY SO MUCH SNOW FALLS ON MOUNT BAKER

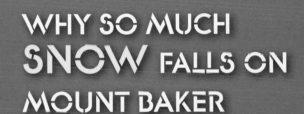

clouds drop their moisture as snow

moist air from ocean

condensation
Condensation is when matter changes from gas to a liquid.

The Windiest The top of Mount Washington in New Hampshire is a windy place. On an average day, the wind blows at about 56 kilometers per hour (35 miles per hour). That's about as fast as a car drives on a city street. An anemometer at the top of the mountain recorded a wind speed of 372 kilometers per hour (231 miles per hour). That's as fast as winds blow inside the most powerful of tornadoes!

Scientists measure parts of weather at the Summit Station of Mount Washington.

The winds at the top of Mount Washington are always strong. This person is leaning into the wind. It is so strong that it holds him up and keeps him from falling to the ground.

The Most Tornadoes There's an area in the middle of the United States called "Tornado Alley." This area gets more tornadoes than any other place. Why here? The area is open and flat. During spring, cold **air masses** from the Arctic collide with warm air masses from the Gulf of Mexico. As they collide, violent storms form at **fronts** . Many of these storms produce tornadoes. The central United States has an average of about 1,200 tornadoes each year.

TECHTREK
myNGconnect.com

Digital Library

The way air masses move cause more tornadoes to occur in this part of the United States than any other.

air mass

An **air mass** is a huge body of air that takes on the temperature and humidity of the land over which it forms.

front

A **front** is the place where two different air masses meet.

The Foggiest The people who live near Cape Disappointment don't see the sun much. It's one of the foggiest places in the United States. Thick fog covers the area about 107 days out of each year. Why is the area so foggy? It sits on the coast of Washington State. **Evaporation** from the ocean makes the air above it have high **humidity**. In summer and early fall, this air passes over a cold ocean current as it moves toward the shore. The water vapor in the air cools. It condenses into water droplets. This creates clouds of fog that move onshore.

Many of the foggiest places are along coasts, as along this coastline in Olympic National Park, Washington.

evaporation
Evaporation is when matter changes from a liquid to a gas.

humidity
Humidity is water vapor in the air.

266

The Wettest   The United States doesn't have the driest spot in the world. But it has one of the wettest. Mount Wai'ale'ale in Hawai'i is the rainiest place in the United States. It gets 12 meters (almost 40 feet) of rain each year. That's almost as tall as a four-story building! Cherrapunji, India, probably holds the world record. In 1974, 24.5 meters (about 80 feet) of rain fell. That's more than twice the total for Mount Wai'ale'ale.

At Mount Wai'ale'ale in Hawai'i, clouds full of moisture rise over mountains. Water in the clouds condenses as the clouds rise, and they drop huge amounts of rain.

CHAPTER 6

SHARE AND COMPARE

Turn and Talk How do landforms and the atmosphere work together to cause extreme weather? Form a complete answer to this question together with a partner.

Read Select two pages in this section. Practice reading the pages. Then read them aloud to a partner. Talk about why the pages are interesting.

Write Write a conclusion that tells the important ideas about what you have learned about extreme weather. State what you think is the Big Idea of this section. Share what you wrote with a classmate. Compare your conclusions. Did your classmate recall that mountains can cause air masses to drop water as snow?

Draw Choose one of the locations of extreme weather. Draw how people might be dressed and an activity they might do in this extreme weather. Put your drawings together with those of your classmates to make a display for "Living in Extreme Weather."

Glossary

A

abrasion (a-BRĀ-zhun)
Abrasion is the scraping away of materials. (p. 152)

air mass (AIR MAS)
An air mass is a large body of air that takes on the temperature and humidity of the land over which it forms. (p. 246)

atmosphere (AT-mus-fear)
Atmosphere is the layer of gases around Earth. (p. 226)

axis (AK-sis)
An axis is an imaginary line around which Earth spins. (p. 12)

C

condensation
(kon-den-SĀ-shun)
Condensation is when matter changes from a gas to a liquid. (p. 238)

conservation
(kon-suhr-VĀ-shun)
Conservation is the protection and care of natural resources. (p. 121)

D

deposition (de-pō-ZI-zhun)
Deposition is the laying down of sediment and rock in a new place. (p. 159)

E

earthquake (URTH-kwāk)
An earthquake is the shaking of the ground usually caused by the movement of Earth's crust or by a volcano. (p. 197)

erosion (ē-RŌ-zhun)
Erosion is the picking up and moving of sediment to a new place. (p. 158)

evaporation
(ē-vap-u-RĀ-shun)
Evaporation is when matter changes from a liquid to a gas. (p. 238)

F

fault (FAWLT)
A fault is a crack in Earth's crust where slabs of rock can slip past, move away from, or push against each other. (p. 196)

fossil fuel (FOS-ul FYŪ-ul)
A fossil fuel is a source of energy that formed from the remains of things that lived millions of years ago. (p. 110)

front (FRUNT)
A front is the place where two different air masses meet. (p. 248)

G

grain (GRĀN)
Grains are small mineral or rock pieces. (p. 61)

gravity (GRA-vi-tē)
Gravity is a force that pulls objects toward each other. (p. 18)

H

humidity (hyū-MID-i-tē)
Humidity is the water vapor in the air. (p. 237)

I

igneous rock (IG-nē-us ROK)
Igneous rock is rock that forms when melted rock cools and hardens. (p. 70)

Wind blew sediment against the surfaces of these rocks and caused abrasion.

L

landform (LAND-form)
A landform is a natural feature on Earth's surface. (p. 146)

latitude (LA-ti-tūd)
Latitude is how far north or south of the Equator a place is. (p. 21)

lava (LAH-vah)
Lava is melted rock that flows from a volcano onto Earth's surface. (p. 200)

M

magma (MAG-muh)
Magma is melted rock beneath Earth's surface. (p. 200)

metamorphic rock (met-a-MOR-fik ROK)
Metamorphic rock is rock that has been changed by heat or pressure. (p. 76)

mineral (MIN-u-ruhl)
A mineral is solid, nonliving material that forms in nature. (p. 59)

N

natural resources (NA-chur-ul RĒ-sors-es)
Natural resources are materials that are found on Earth that people use. (p. 98)

nonrenewable resources (non-rē-NŪ-uh-bul RĒ-sors-es)
Nonrenewable resources are materials that cannot be replaced quickly enough to keep from running out. (p. 98)

O

ore (ŌR)
Ore is rock that contains metal. (p. 108)

P

phase (FĀS)
A phase is a lighted part of the moon as it appears from Earth. (p. 35)

plate (PLĀT)
A plate is a large section of Earth's crust and outer mantle that slowly moves. (p. 195)

property (PROP-ur-tē)
A property is something about an object you can observe with your senses. (p. 60)

R

renewable resources (rē-NŪ-uh-bul RĒ-sors-es)
Renewable resources are materials that are always being replaced and will not run out. (p. 98)

revolve (re-VAWLV)
To revolve is to move around another object. (p. 18)

rotate (RO-tāt)
To rotate is to spin around. (p. 12)

S

sediment (SED-ah-mint)
Sediment is material that comes from the weathering of rock. (p. 152)

sedimentary rock (sed-ah-MIN-tair-ē ROK)
Sedimentary rock is rock that forms when small rock pieces and other materials settle and then get squeezed or cemented together. (p. 72)

V

volcano (vol-KĀ-nō)
A volcano is a mountain or hill on Earth's crust through which lava, gas, and ashes can erupt. (p. 200)

W

weathering (WE-thur-ing)
Weathering is the breaking apart, wearing away, or dissolving of rock. (p. 148)

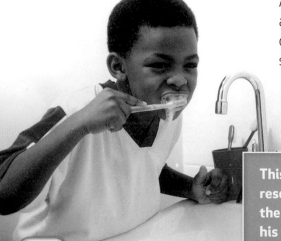

This boy conserves natural resources by turning off the water while brushing his teeth.

Index

A

Abrasion, 145, 152, 174, 184

Acid, minerals' reaction with, 67

Africa, 160

Air
moves as wind, 236
pollution, 124–125
as renewable natural resource, 104–105, 118

Air mass
conclusion, 256
continental, 253
definition of, 225, 246, 265
maritime, 253
origins of, 252–253
polar, 253
tropical, 253
vocabulary review, 256
weather and, 247

Air pressure
differs from place to place, 235
measuring, 231
as part of weather, 231
weight of air and, 234

Al-Battani, 45

Aluminum, 109, 123

Analyze, 129, 209, 237

Andes Mountains, 71

Antarctica, 262

Apply, 41, 69, 105, 117, 257

Aryabbata, 46

Ash, volcanic, 202–203

Astronaut, 30–31

Astronomer
Arab, 45
Chinese, 46
cultural, 42–43
Greek, 45
Indian, 46
Mayan, 47
modern, 50–51

Atacama Desert, 260

Atlantic Ocean, 161

Atmosphere
air and, 226–229
big idea review, 257
clouds in, 241, 244
conclusion, 256
definition of, 224, 226, 260
gases in, 228
layers of, 228, diagram, 229
vocabulary review, 256
weather occurs in, 226, 229
wind and, 226

Axis
definition of, 8, 12, 46
of Earth, 8, 12, 19–20, 32
of moon, 32
tilt of, 19–20, 22–25
vocabulary review, 40

B

Bacteria, 115

Barometer, 231

Become an Expert
The Earth-Moon-Sun System: How Knowledge Grows, 44–52
Extreme Weather, 260–268
The Grand Canyon: History Written in Rock, 84–92
The Hawaiian Islands: Formed by Volcanoes, 212–220
Making Blue Jeans: Resources on the Move, 132–140
Yosemite Valley: Shaped by Weathering and Erosion, 178–188

Before You Move On, 11, 17, 25, 31, 35, 61, 69, 77, 99, 105, 113, 117, 125, 147, 151, 157, 159, 165, 171, 195, 199, 203, 205, 229, 237, 245, 253

Big Idea Questions
How Are Rocks Alike and Different? 53–92
How Do Earth and Its Moon Move? 5–52

How Do Slow Processes Change Earth's Surface? 141–188
What Are Renewable and Nonrenewable Resources? 93–140
What Can We Observe About Weather? 221–268
What Changes Do Volcanoes and Earthquakes Cause? 189–220

Big Idea Review, 41, 81, 129, 175, 209, 257

Binoculars, 27–28

Biographical information about scientists. See Earth Science Expert; Meet a Scientist

Blue jeans, 132–139

Bodies of water, changed by erosion and deposition, 163

Bolivia, 172–173

Borneo rain forest, 126–127

C

Calendar, 34, 36, 46

Canyons, 84–91, 144–147

Cape Disappointment, Washington, 266

Carbon dioxide, 105

Careers, science. See Earth Science Expert; Meet a Scientist

Cause and effect, 41, 175, 229, 257

Caves
formation of, 151
sinkholes and, 166

Changes
caused by earthquakes, 198–199, 205
caused by landslides, 204
caused by volcanoes, 200, 202–204
in air pressure, 235
in clouds, 241, 243
in Earth's surface, rapid, 189–220

in Earth's surface, slow, 141–188
in seasons, 20–25
in sky, 14–17, 26, 34–35
in soil, 115
in temperature, 22–25, 232
in weather, 245, 248–251
write about, 81

Characteristics
of Earth, 10–11, 12–13, 18–21, 40
of moon, 10–11, 28–35, 40
of seasons, 22–25
of sun, 10–11, 18

Charts
characteristics of sun, Earth, moon, 11
cloud shapes, 238–239
coal-fired plant, how it works, 111
copper production, 108–109
cotton pod to cotton fibers, 134
earthquake magnitude scale, 198
glaciers of Yosemite Valley, 180
ice wedging, 181
landforms, 147
Moh scale of hardness, 64–65
moon phases, 34
rocks, properties of, 60
recycled items, 123
renewable and nonrenewable resources, 99
precipitation, types of, 233
soil, how it forms, 115
soil, types of, 117
volcanoes, types of, 201
water, uses of in the home, 120
weathering, 148–149

Cherrapunji, India, 267

Cinder cone volcano, 201

Cirrus clouds, 239, 244–245, 250

Cleavage, as property of minerals, 66

Climbing Weather, 254–255

Clock, 36, 38

Clouds
cirrus, 239, 244–245, 250
cumulonimbus, 241, 248
cumulus, 238, 240–241
fog, 242, 266
formation of, 238–240, 244
humidity and, 237
nimbostratus, 243
precipitation, conditions that form, and, 241–243, 245
stratus, 238–239, 242–243, 250
water cycle and, 238
weather and types of, 240–245

Coal, 98–99, 110–111

Coastline, 4, 146–147, 168–169

Cole, Johari, 130–131

Color
as property of minerals, 62
as property of rock, 60, 73, 85, 88
sedimentary rocks and, 73, 85

Colorado River, 84, 155

Compare, 157

Compare and contrast, 41, 81, 159, 165, 209

Composite volcano, 201

Composting, 118

Conclusion, 40, 80, 128, 174, 208, 256

Conclusions, draw, 41, 129, 175, 199, 209, 253, 257

Condensation, 225, 238, 256, 263

Conservation, 97, 121, 128, 139

Contrast, 99, 257

Copernicus, Nicolaus, 48

Copper, 108–109, 123

Cotton
bales of, 136
blue jeans and, 133–137
fossil fuels in transporting, 136
picking of, 135
as renewable resource, 134
seeds, 134–136

Cotton gin, 135

Crater, 29

Crops, 103, 116

Cultural astronomer, 42–43

Culture, 42–43, 44–47

Cumulonimbus clouds, 241, 248

Cumulus clouds, 238, 240–241

Cycles
day and night, 12–17, 40
of moon's appearance, 9, 34–35, 40
water, 100–101, 104

D

Day, 12–15

Define, 81

Delta, 163

Denim, 133, 137–138

Deposition
big idea review, 175
conclusion, 174
definition of, 145, 159, 185
glaciers and, 159, 185–186
gravity and, 161, helpful effects of, 170–171
humans affect, 169
by ice, 165
vocabulary review, 174
by water, 162–163, 168–169, 187
by wind, 161, 187
write about, 175

Describe, 81, 129, 175, 209, 257

Desert, 146, 260

Diagrams
atmosphere, layers of Earth's, 229

caves, formation of, 151
cold front, 248
constellations, 26
Earth's layers, 194
Earth's revolution, 18
Grand Canyon, rock layers of, 89
hanging valleys, 183
hot spots, formation of island chains, 214
hot spots, prediction of activity, 215
moon, revolution of, 33
moraines, formation of, 185
Mount Baker, why so much snow falls on, 263
oil and natural gas, 113
seamounts, formation of, 218
U-shaped valleys, 182
warm front, 250
water cycle, 100–101

Diamonds and Their Uses, 78–79

Digital Library, 5, 7, 13, 17, 21, 23, 30, 34, 43, 45, 47, 53, 55, 59, 74, 83, 85, 89, 93, 95, 98, 103, 117, 118, 130, 131, 133, 135, 137, 141, 143, 151, 164, 172, 176, 177, 179, 186, 189, 191, 201, 210, 211, 213, 217, 221, 223, 231, 233, 259, 261, 265

E

Earth. *See also* Earth's structure; Earth's surface, changes to
ability to support life of, 11
axis of, 12, 19–20, 32
big idea review, 41
changes to surface of, rapid, 189–220
changes to surface of, slow, 141–188
characteristics of, 11
compared to moon, sun, 10–11
conclusion, 40
core of, 194
crust of, 194–195
day created by spin of, 12–15

distance from moon, sun, 11
gravity and, 9, 18–19, 32, 49
latitude and, 9, 21
mantle of, 194–195
movement of, 8, 12–17, 18–26, 32, 40, 44, 46, 48
night created by spin of, 12–13, 16–17
orbit of defines a year, 8, 18, 47
as part of solar system, 10, 40
position of, 26
revolution of, 8, 18, 22–26, 32, 40, 44, 48
rotation of, 8, 12–17, 32, 40, 46, 48
seasons caused by movement and position of, 19–25
size of, 10–11
structure of, 194–195
surface of, 11
telling time and, 36–37
vocabulary review, 40
weathering of surface, 148–151
write about, 41

Earthquake
definition of, 192, 197, 212
vocabulary review, 208

Earthquakes
big idea review, 209
causes of, 196–197
changes to Earth's surface by, 198
conclusion, 208
damage caused by, 198–199
faults and, 196
in Japan, 206–207
landslides caused by, 205
magnitude scale, 198, *chart, 198*
plates and, 196–197
preparing for, 206–207
seismographs, 198–199

Earth Science Expert
Cole, Johari, 130–131
cultural astronomer, 42–43
Fleming, Edward, 82–83

glaciologist, 176–177
Holbrook, Jarita, 42–43
Johnson, Veronica, 258–259
Mather, Tamsin, 210–211
meteorologist, 258–259
naturalist, 130–131
sculptor, 82–83
Thompson, Lonnie, 176–177
volcanologist, 210–211

Earth's structure
core, 194
crust, 194–195
mantle, 194–195, 214
plates, 195–197

Earth's surface
abrasion, 145, 152, 184
big idea review, 175, 209
changes in, 144–188, 189–220
conclusion, 174, 209
deposition, 145, 159, 161, 163, 165, 174, 185, 187
earthquakes, 196–199, 206–207, 208
erosion, 145, 158, 160–164, 174, 183, 187
landslides, 167, 204–205, 208
vocabulary review, 174, 208
volcanoes, 193, 200–203, 206–207, 208, 212–219
weathering, 144, 148–157, 174, 181, 187
write about, 175, 209

Eclipse, solar, 42–43

eEdition. See Student eEdition

Electricity, 94–95

Emperor Seamounts, 218–219

Enrichment activities, 5, 7, 18, 53, 55, 70, 72, 76, 93, 95, 101, 141, 143, 152, 154, 189, 191, 199, 221, 223, 236

Environmental problems, solving
composting, 118
conservation, 97, 121, 139
recycle, 123, 128–129

Environment, humans change
in harmful ways, 118, 120, 124, 127
in helpful ways, 102–103, 119, 121, 125

Erosion,
big idea review, 75, 175
conclusion, 174
definition of, 145, 158, 183
by glaciers, 159, 164–165, 178, 180, 182–184
gravity and, 158
hanging valleys formed by, 183
helpful effects of, 170–171
humans affect, 119
humans effected by, 166–171
by ice, 158, 164–165
plants control, 103
of seamounts, 218
soil, 103, 119
vocabulary review, 174
by water, 103, 158, 162–163, 168–169
by wind, 158, 160–161
Yosemite Valley shaped by, 183, 187

Erratics, 178

Evaluate, 31, 125, 147, 175

Evaporation, 224, 238, 256, 266

Explain, 81, 129, 171, 175, 203, 209

Fall, 23

Farming
environment and, 119, 127
no-till, 119
organic, 130–131

Fault, 192, 196, 209, 212, 213

Fleming, Edward, 82–83

Flood plains, 171

Fog, 242, 266

Forests, 102, 118

Fossil fuel
air pollution and, 124–125
in Borneo rain forest, 127
coal, 98–99, 110–111
definition of, 97, 110, 136
natural gas, 98, 112–113
oil, 99, 112–113
transporting cotton and, 136
vocabulary review, 128

Fossils
marine, 74, 86
terrestrial, 74

Front(s), 225, 248, 256, 265. See also Weather fronts

Galileo, 48

Gas. See Natural gas; Oil

Gems, 78

Generalize, 17, 81

Geo-archaeologist, 4

Glacial polish, 184

Glaciers
deposition of soil, 170, 185
glacial polish, 184
picking up and moving rocks, 159, 164, 186
shrinking, 172–173, 176
as water supply, 173
in Yosemite Valley, 178, 180–187

Glaciologist, 176–177

Glossary, EM1–EM2

Goodman, Beverly, 4

Grain, 56, 61, 70–71, 80, 89

Grain, in rock, 56, 61, 70–71, 89

Grand Canyon, 84–92, 155

Gravity
definition of, 9, 18, 49
deposition and, 161
of Earth, 9, 32, 49
erosion and, 158
landslide and, 167, 205
of moon, 30, 32
of sun, 18–19, 49
vocabulary review, 40

Hardness
Moh scale of, 64–65
as property of minerals, 64–65, 78

Hawaiian Islands, 203, 212–219

Hills, 146

History
of Earth, through fossils, 74–75, 86, 88, 91
Grand Canyon and, 84–91
telling time and, 36–37
of understanding Earth-moon-sun system, 44–51
Yosemite Valley and, 178–187

Holbrook, Jarita, 42–43

Hot spots, 214–215, 218–219

Hour, 13, 38

Humidity
definition of, 224, 237, 266
measuring, 231
as part of weather, 231
vocabulary review, 256
weather predicting and, 237

Humus, 165

Ice, 148, 164–165, 181

Ice wedging, 181

Identify, 41, 209

Igneous rock
changed into metamorphic rock by heat and pressure, 77
definition of, 57, 70, 90–91
formation of, 70–71, 80, 90
in Grand Canyon, 90
vocabulary review, 80

Infer, 11, 25, 61, 77, 81, 151, 175, 195, 205, 209

Interpret diagrams, 81

Iron, 109

Japan, 206–207

Johnson, Veronica, 258–259

Hand lens, 68

Hanging valleys, 183

K

Kepler, Johannes, 49

L

Landform
definition of, 144, 146, 178
vocabulary review, 256

Landforms
canyons, 144–147
changed by erosion, 178, 183–184
changed by weathering, 154–155, 168–169, 181, 183
coastline, 4, 146–147, 168–169
deserts, 146, 160
examples of, *chart, 147*
formed by glaciers, 165, 185
formed by volcanoes, 203, 212–213
hills, 146
mountains, 146–147, 172, 195, 218–219
plains, 146–147
river valleys, 146–147, 171
types of, *chart, 147*
in Yosemite Valley, 179

Landslides
big idea review, 209
causes of, 167, 204–205
changes to Earth's surface by, 167
conclusion, 208
damage caused by, 167
vocabulary review, 208

Latitude
big idea review, 40
definition of, 9, 21
seasons and, 21
vocabulary review, 40

Lava, 29, 70, 193, 200, 209, 213, 216–217

Layers, of Earth's surface.
See Earth's structure

Leap year, 39

Libya, 261

Light
of moon, 33, 46
of sun, 12–15, 22–25, 104

Limestone, 73, 76, 107, 150–151, 166

Living on the Edge in Japan, 206–207

Lunar rover, 31

Luster, as property of minerals, 67, 78

Lyra star pattern, 26

M

Magma, 70–71, 78, 90, 193, 200, 209, 214

Magnetism, as a property of minerals, 66

Magnitude scale, earthquake, 198

Make connections, 113

Make judgments, 129

Maps
air masses, 247, 252–253, 265
equator, 21
globe, 45
Hawaiian Islands, 213
Sahara, 160
weather, 252–253, 255

Mather, Tamsin, 210–211

Meet a Scientist
geo-archaeologist, 4
Goodman, Beverly, 4

Metals, 108–109

Metamorphic rock
big idea review, 81
conclusion, 80
definition of, 57, 76, 90–91
formation of, 76–77, 80
in Grand Canyon, 90
vocabulary review, 80

Meteorologist, 258–259

Microscope, 68

Mineral
definition of, 56, 59, 86
vocabulary review, 80

Minerals
acid, reaction to, 67
in Borneo rain forest, 127
cleavage, 66
color, 62
as gems, 78–79
hardness, 64–65
luster, 67

magnetism, 66
as natural resource, 122–123
properties of, 62–67, 80
in rocks, 69–70, 149–150
sedimentary rock and, 90
streak, 63

Minute, time unit, 38

Moh scale of hardness, 64–65

Month, 32–35

Moon
ability to support life of, 11, 29
axis of, 32
big idea review, 41
characteristics of, 11, 29, 31
compared to Earth, sun, 10–11
conclusion, 40
craters, 29
daytime and, 28
distance from Earth, 11
gravity of, 30, 32
light of, 33, 46
month and, 32–35
motion of, 6, 32–35, 40, 44
observing the, 10, 28–31
phases of, 9, 34–35, 40
predicting future appearance of, 34
relationship to Earth of, 6, 16–17, 32–35
revolution of around Earth, 32–35, 40, 44
rotation of, 6, 32–33, 40
size of, 10–11
solar eclipse and, 42
surface of, 11, 29
temperatures on, 30
vocabulary review, 40

Moraines, 165, 185

Mountains
formation of, 195
as landform, 146–147
Mount Baker, 263
Mount Rainier, 254–255
Mount Vesuvius, 203
Mount Waialeale, 267
Mount Washington, 264

Tuni Condoriri mountain range, 172

My Science Notebook
draw, 52, 92, 140, 188, 220, 268
write, 52, 92, 140, 188, 220, 268
write about
Earth's movement, 41
Earth's surface, 175, 209
how rocks form and change, 81
resources, 129
weather, 257

N

Natural gas, 98, 112–113

Naturalist, 130–131

Natural resources. *See also* Nonrenewable resources; Renewable resources
big idea review, 129
conclusion, 128
conservation of, 97, 121, 123, 139
definition of, 96, 98, 132
environmental change and, 118, 127
humans' dependence upon, 100, 104–105, 116, 118, 132–133
humans' effect on, 118–121, 123–125, 127
nonrenewable, 98–99, 118
recycling of, 123, 128–129
renewable, 94–95, 98–105, 118
vocabulary review, 128
write about, 129

New Hampshire, 264

Newton, Sir Isaac, 49

Night, 12–13, 16–17, 28

Nimbostratus clouds, 243

Nitrogen, 228

Nonrenewable resources
big idea review, 129
coal, 98–99, 110–111
conclusion, 128
definition of, 96, 98, 138
fossil fuels, 110–113,

124–125, 127, 128, 136
metals, 108–109
natural gas, 98, 112–113
oil, 99, 112–113
ores, 97, 108–109,
122–123, 128
recycling, 123
rocks, 106–108,
122–123
soil, 114–117, 119, 128
vocabulary review, 128

Northern Hemisphere,
22–25

North Pole, 12

No-till farming, 119

O

Objects in sky
moon, 10, 16–17, 28–31
patterns of movement
of, 14–17, 32–33
stars, 10, 16–17, 26–27
sun, 10, 14–15, 17

Oil, 99, 112–113, *See*
"Nonrenewable
Resources, oil"

Oil rig, 113

Orblt, 18, 32, 49

Ore, 97, 108–109,
122–123, 128

Orion, 26

Oxygen, 105, 116, 126, 228

P

Pacific Ocean, 214

**Patterns of moon's
appearance.** *See* Phases,
of moon

Peat, 111

Phase
definition of, 9, 35, 48
vocabulary review, 40

Phases
of moon, 9, 34–35, 40
of Venus, 48

Plains, 146–147

Plants
weathering and, 148,
156–157

Plate, 192, 195, 196–197,
209, 214–215

Pollution, 120, 124–125

Precipitation
clouds and, 233, 241,
243, 245
in deserts, 260
measuring, 231
as part of weather, 231
types of, *chart*, 233

Predict, 35, 245, 257

**Processes, change Earth's
surface**
rapid, 189–220
slow, 144–188

Products
of animals, 99, 116, 132
of plants, 99, 102–103,
116, 132–139
reusable, 123

Properties
of minerals, 62–67, 80
of rocks, 60–61, 80, 85

Property
definition of, 56, 60, 85
vocabulary review, 80

Prospect Creek, Alaska,
262

R

Rain forest, 126–127

Recycle, 123, 128–129

Renewable resources
air, 104–105, 118,
See "Nonrenewable
Resources, oil"
big idea review, 129
blue jeans and, 132–139
conclusion, 128
crops, 103
definition of, 96, 98, 134
fish, 99
forests, 99, 102, 118
living things, 102–103,
126–127
plants, 98, 102–103
sunlight, 104, 125
vocabulary review, 128
water, 98, 100–101, 118,
120–121, 125, 128

wind, 94, 125, 128

**Resources in the Borneo
Rain Forest,** 126–127

Resources, natural. *See*
Natural resources

Restate, 41, 129, 175, 257

Revolution
of Earth, 8, 18, 22–26,
32, 40, 44, 48
of moon, 32–35, 40, 44

Revolve, 8, 18, 40, 44

River valleys, 146–147, 171

Rocks,
big idea review, 81
compare and contrast,
53–92
composed of minerals,
59, 61, 80
conclusion, 80
definition of, 59
grains in, 56, 61, 70–71,
89, *See* "Grain, in rock"
in Grand Canyon, 84–89
igneous, 57, 70–71, 77,
90–91
metamorphic, 57, 76–77,
80, 90–91
minerals and, 149–150
as natural resources,
106–109, 122–123
ores, 108–109
properties of, 60–61,
80, 85
sedimentary, 57, 72–77,
84–90
types of, 70–77, 80
vocabulary review, 80
weathering of, 148–157
write about, 81

Rocky Mountains, 82

Rotate, 8, 12, 40, 46

Rotation of Earth
axis and, 12, 46
Copernicus, Nicolaus,
and, 48
day and, 8, 12–15
night and, 8, 12–13,
16–17
24 hour cycle and, 13

Rotation of moon, 32–33,
40

S

Sahara Desert, 160, 261

Sandstone, 72–73, 77

Sandstorm, 160

Science in a Snap!
Air Takes Up Space, 227
Buried Treasure, 75
Night and Day, 12
Pressure Buildup, 197
Sun Power, 104
Wearing Away, 153

Science Vocabulary, 8–9,
56–57, 96–97, 144–145,
192–193, 224–225
abrasion, 145, 152,
174, 184
air mass, 225, 246,
256, 265
atmosphere, 224, 226,
256, 260
axis, 8, 12, 40, 46
condensation, 225, 238,
256, 263
conservation, 97, 121,
128, 139
deposition, 145, 159,
174, 185
earthquake, 192, 197,
208, 212
erosion, 145, 158,
174, 183
evaporation, 224, 238,
256, 266
fault, 192, 196, 208,
212, 213
fossil fuel, 97, 110,
128, 136
front, 225, 248,
256, 265
grain, 56, 61, 80, 89
gravity, 9, 18, 40, 49
humidity, 224, 237,
256, 266
igneous rock, 57, 70,
80, 90, 91
landform, 144, 146,
174, 178
latitude, 9, 21, 40
lava, 193, 200,
208, 213

magma, 193, 200, 208, 214

metamorphic rock, 57, 76, 90, 91

mineral, 56, 59, 80, 86

natural resources, 96, 98, 128, 132

nonrenewable resources, 96, 98, 128, 138

ore, 97, 108, 128

phase, 9, 35, 40, 48

plate, 192, 195, 208, 214

property, 56, 60, 80, 85

renewable resources, 96, 98, 128, 134

revolve, 8, 18, 40, 44

rotate, 8, 12, 40, 46

sediment, 144, 152, 174, 184

sedimentary rock, 57, 72, 80, 84

volcano, 193, 200, 208, 212

weathering, 144, 148, 174, 181

Scientists. *See* Earth Science Expert; Meet a Scientist

Sculptor, 82–83

Seamount, 215, 218–219

Seasons

causes of, 19–25

changes in, 20–25

Earth's orbit and, 19–20

Earth's tilt on axis and, 19–20, 22–25

fall, 23

latitude and, 21

length of daylight varies with, 22–25

shadows and, 22–25, 47

spring, 25, 47

stars and, 26, 40

summer, 22

temperatures and, 22–25

winter, 24

Second, time unit, 38

Sediment, 72, 144, 152, 154–155, 158–159, 162–163, 165, 168, 174, 184–185, 187

Sedimentary rock

changed into metamorphic rock by heat and pressure, 76–77, 90

color of, 73

definition of, 57, 72, 84

formation of, 72, 80

in fossils, 74–75, 88

in Grand Canyon, 84–89

vocabulary review, 80

Seismographs, 198–199

Sequence, 129

Shadows

seasons and, 22–25

sun, daily motion of and, 14–15

time of day and, 14–15

Shape of moon, visible. *See* Moon, phases of

Share and Compare, 52, 92, 140, 188, 220, 268

Shield volcano, 201

Shrinking Glaciers in South America, 172–173

Sinkhole, 166

Sky

day, changes in, 14–15

moon appears to move across, 16–17, 40, 44

night, changes in, 16–17, 26, 34–35

objects in, 10, 14–17, 26–35

star patterns, 16–17, 26

stars appear to move across, 16–17, 40, 44

sun appears to move across, 14–15, 40, 44, 48

Soil

components of, 115

enriched by volcanoes, 203

erosion of, 103, 119

as nonrenewable resource, 114–117, 128

types of, *chart,* 117

Solar system, 10, 40

South Pole

coldest temperature at, 262

Earth's axis and, 12

Spacecraft, 28, 50

Spring, 25

Stars

navigating by, 42

observing, 10, 26–27

patterns of, 16–17, 26

seasons and, 26, 40

Stratus clouds, 238–239, 242–243, 250

Streak, as property of minerals, 63

Student eEdition, 5, 7, 43, 45, 53, 55, 83, 85, 93, 95, 131, 133, 141, 143, 177, 179, 189, 191, 211, 213, 221, 223, 259, 261

Summer, 22

Sun

ability to support life of, 11

characteristics of, 11

compared to Earth, moon, 10–11

day sky and, 14–15

dependence of living things on, 104

distance from Earth, 11

Earth's movement around, 8, 22–26, 32, 40, 44, 48

eclipse of, 42–43

as energy source, 104

gravity of, 18–19, 49

light of, 33

as natural resource, 104, 125

seasons and, 22–25

size of, 10–11

telling time and, 36–37

Technology

astronomy and, 27, 50–51

Tech Trek

digital library, 5, 7, 13, 17, 21, 23, 30, 34, 43, 45, 47, 53, 55, 59, 74, 82, 83, 85, 89, 93, 95, 98, 103, 117, 118, 130, 131, 133, 135, 137, 141, 143, 151, 164, 172, 176, 177, 179, 186, 189, 191, 201, 210, 211, 213, 217, 221, 223, 231, 233, 259, 261, 265

enrichment activities, 5, 7, 18, 53, 55, 70, 72, 76, 93, 95, 99, 141, 143, 152, 154, 189, 191, 199, 221, 223, 236

student eEdition, 5, 7, 43, 45, 53, 55, 83, 85, 93, 95, 131, 133, 141, 143, 177, 179, 189, 191, 211, 213, 221, 223, 259, 261

vocabulary games, 5, 7, 9, 53, 55, 57, 93, 95, 97, 141, 143, 145, 189, 191, 193, 221, 223, 225

Telescopes, 27–28, 48, 50–51

Telling Time: Yesterday and Today, 36–39

Temperature

changes in, 232

measuring, 231

as part of weather, 231

physical weathering and, 157

seasonal, 20–25, *See* "Seasons, temperature and"

Thinking skills

analyze, 129, 209, 237

apply, 41, 69, 105, 117, 257

cause and effect, 41, 175, 229, 257

compare, 157

compare and contrast, 41, 81, 159, 165, 209

conclusions, draw, 41, 129, 175, 199, 209, 253, 257

contrast, 99, 257

define, 81

describe, 81, 129, 175, 209, 257

evaluate, 31, 125, 147, 175

explain, 81, 129, 171, 175, 203, 209

generalize, 17, 81

identify, 41, 209

infer, 11, 25, 61, 77, 81, 151, 175, 195, 205, 209

interpret diagrams, 81

make connections, 113

make judgments, 129

predict, 35, 245, 257

restate, 41, 129, 175, 257

sequence, 129

Thompson, Lonnie, 176–177

Till, 165, 185

Time, telling and organizing
calendar, 34, 36, 46
clock, 36, 38
day, 38
Earth and, 36–37
hour, 38
leap year, 39
minute, 38
month, 32–35
moon and, 32–35, 36
second, 68
sun and, 36–37, 39

Tornado Alley, 265

Tsunami, 4

Tuni Condoriri mountain range, 172

U

Units of time. *See* Time, telling and organizing

U-Shaped valleys, 182

V

Valleys
hanging, 183
U-Shaped, 182

Venus, 48

Vocabulary games, *See* "Tech Trek" 5, 7, 9, 53, 55, 57, 93, 95, 97, 141, 143, 145, 189, 191, 193, 221, 223, 225

Vocabulary review, 40, 80, 128, 174, 208, 256

Volcano
definition of, 193, 200, 212
vocabulary review, 208

Volcanoes
active, 213, 200, 204
ash, 202–203
big idea review, 209
cinder cone, 201
composite, 201
conclusion, 208
damage caused by, 202
extinct, 213
in Hawaiian Islands, 212–219
in Iceland, 200
in Japan, 206–207
landform created by, 203
landslides caused by, 204
lava and, 200–201, 213
magma and, 70, 200
Mount Vesuvius, 203
Mount St. Helens, 204
preparing for, 206–207
shield, 201, 213
soil enriched by, 203
types of, *chart, 201*

Volcanologist, 210–211

W

Washington, state, 254, 263, 266

Water
condensation, 100
conservation of, 121
cycle, 100–101, 104
deposition by, 162–163, 168–169, 187
erosion by, 148, 150–151, 162–163, 168–169
evaporation, 100
forms of, in the air, 237–243, 249, 251
humans benefit from uses of, 120
as natural resource, 98, 100–101, 118, 120–121, 125, 128
pollution of, 120
precipitation, 101
weathering by, 148, 154–155, 187

Water vapor, 228, 237, 238–239, 260, 266

Weather
air pressure, 231, 234–235, 256
big idea review, 257
conclusion, 256
definition of, 230
extreme, 260–267
humidity, 231, 237, 256
measuring, 231, *chart, 256*
Mount Rainier and, 254–255
observing, 221–268
parts of, 230–237, 256
precipitation, 231, 233, 256
predicting, 258–259
temperature, 231–232, 256
vocabulary review, 256
wind, 226, 231, 256
write about, 257

Weather fronts
cold, 248–249
definition of, 225, 248, 265
rain and, 251
storms and, 249
in United States, 252–253
warm, 250–251

Weathering
definition of, 144, 148, 181
vocabulary review, 174

Weathering, physical
abrasion, 145, 152, 184
big idea review, 175
by chemicals in water, 148, 150–151
conclusion, 174
Earth's surface and, 148–151
by ice, 148, 181
landslides caused by, 205
people affected by, 166–171
by plants' growth, 148, 156–157
soil formation and, 115
by temperature change, 157
by water, 148, 154–155, 187
by wind, 148, 152–153, 187
Yosemite Valley shaped by, 181, 187

Weather patterns
air masses and, 253
west to east movement of, *map, 252–253*

What Is Earth Science? 2–3

Wind
air pressure causes, 236
cold fronts and, 248
cumulonimbus clouds and, 241
deposition by, 161, 187
erosion by, 160–161
measuring, 231
as natural resource, 94, 125
as part of weather, 231
weathering and, 148, 152–153, 187

Wind farm, 94–95

Winter, 24

Write about
Earth's movement, 41
Earth's surface, 175, 209
how rocks form and change, 81
resources, 129
weather, 257

Y

Year, 8, 18, 47

Yosemite Valley, 178–187
abrasion in, 184
erratics in, 186–187
glacial polish in, 184
glaciers of, 178, 180–187
hanging valleys in, 183
ice wedging in, 181
landforms of, 178
moraines in, 185
shaped by erosion and deposition, 183, 187
shaped by ice, 181
shaped by weathering, 181, 187
U-Shaped valleys in, 182

Credits

Front Matter **About The Cover** (bg) Kevin M. Law/Alamy Images. (t inset) Kevin M. Law/Alamy Images. (b inset) Tim Fitzharris/Minden Pictures/National Geographic Image Collection. **ii–iii** (bg) Kenneth Garrett/National Geographic Image Collection. **iv–v** Michael T. Sedam/Corbis Premium RF/Alamy Images. **vi–vii** (bg) John Eascott and Yva Momatiuk/National Geographic Image Collection. **vii** (t) Images & Stories/Alamy Images. **viii–ix** (bg) Digital Vision/Getty Images. **ix** (t) Jay Dickman/National Geographic Image Collection. **x–1** Frans Lemmens/Getty Images. **2** (t) DigitalStock/Corbis. (c) John Burcham/National Geographic Image Collection. (b) Glen Allison/Photodisc/Getty Images. **2–3** (bg) D Woodfall/ImageState/Panoramic Images. **3** (t) David Boyer/National Geographic Image Collection. (c) Steve and Donna O'Meara/National Geographic Image Collection. (b) Digital Vision/Getty Images. **4** (t) Beverly Goodman. (b) Amir Yurman/The Leon Recanati Institute for Maritime Studies.

Chapter 1 **5, 6–7** DigitalStock/Corbis. **8** (tl) Grant Dixon/Hedgehog House/Minden Pictures/National Geographic Image Collection. (tr) Torsten Stahlberg/iStockphoto. (c) Stefano Maccari/Shutterstock. **9** (t) rotofrank/iStockphoto. (c) Paul A. Souders/Corbis. (b) max romeo/Shutterstock. **10** (l) Artville. **10–11** (bg) rotofrank/iStockphoto. **11** (t) Creatas/Jupiterimages. (c) PhotoDisc/Getty Images. (b) DigitalStock/Corbis. **12** (tr) Stefano Maccari/Shutterstock. **12–13** (bg) Grant Dixon/Hedgehog House/Minden Pictures/National Geographic Image Collection. **13** (inset) Torsten Stahlberg/iStockphoto. **14–15** Steve Casimiro/The Image Bank/Getty Images. **16** (inset) Bill Hatcher/National Geographic Image Collection. **16–17** (bg) Don Smith/Getty Images. **17** (inset) David Nunuk/Photo Researchers, Inc. **18–19** (r) Kim Kyung-Hoon/Reuters/Corbis. **20–21** Paul A. Souders/Corbis. **022** peter dazeley/Alamy Images. **23** John Warburton-Lee Photography/Alamy Images. **24** (inset) Dave Reede/First Light/Getty Images. **24–25** Mikhail Tolstoy/Alamy Images. **26–27** Steve Cole/Photodisc/Alamy Images. **27** (inset) Robert Sisson/National Geographic Image Collection. **28–29** (bg) DigitalStock/Corbis. **29** (t) Roger Ressmeyer/Corbis. (b) Stockbyte/Getty Images. **30–31** Stockbyte/Getty Images. **31** (inset) NASA Human Space Flight Gallery. **32–33** max romeo/Shutterstock. **34** Midhat Becar/iStockphoto. **36** (l) Alexander Maksimenko/iStockphoto. **36–37** Peter Adams/Getty Images. **37** (tl) Helene Rogers/Alamy Images. **38** (tl) vrjoyner/Shutterstock. **38–39** Jeremy Baer-Simon/NIST. **39** (tl) Patryk Galka/iStockphoto. **40** (l) Creatas/Jupiterimages. (c) PhotoDisc/Getty Images. (r) DigitalStock/Corbis. **40–41** (bg) D. Nunuk/Photo Researchers, Inc. **41** (inset) Maria Stenzel/National Geographic Image Collection. **42** David E. Franck/Franck Fotos, Inc. **43** (b) Images & Stories/Alamy Images. (inset) Digital Vision/Getty Images. **44–45** Dougal Waters/Digital Vision/Getty Images. **45** (r) Topham/The Image Works, Inc. **46** (tr) Donalee Houston. (b) William Ju/Shutterstock. **47** (t) hannahgleg/iStockphoto. (b) Stephen Beyer/National Geographic Image Collection. **48** (c) Erich Lessing/Art Resource, Inc. **48** (inset) Lea and-Leon Huens/National Geographic Image Collection. **49** Duncan Walker/iStockphoto. **50–51** T. Abbott and NOAO/AURA/NSF/NOAO. **51** (inset) Roger Ressmeyer/Corbis. **52** (b) Erich Lessing/Art Resource, Inc.

Chapter 2 **53, 54–55** Jerry Dodrill/Aurora/Getty Images. **56** (t) Andrei Merkulov/Shutterstock. (c) Natural Selection Steven Raniszewski/Design Pics Inc./Alamy Images. (b) Edward Kinsman/Photo Researchers, Inc. **57** (t) Jim Sugar/Corbis. (c) Mike Brake/Shutterstock. (bl) Gary Ombler/Dorling Kindersley/Getty Images. (br) Mick Rock/Alamy Images. **58–59** (bg) John Burcham/National Geographic Image Collection. **59** (l) Scientifica/Visuals Unlimited. (c) Andrei Merkulov/Shutterstock. **60** (r1–l) Joel Arem/Photo Researchers, Inc. (r1–r) Bragin Alexey/Shutterstock. (r2–l) Edward Kinsman/Photo Researchers, Inc. (r2–r) Mark Schneider/Visuals Unlimited. (r3–l) Dave King/Dorling Kindersley Ltd. Picture Library. (r3–r) Biophoto Associates/Photo Researchers, Inc. (r4–l) Natural Selection Steven Raniszewski/Design Pics Inc./Alamy Images. (r4–r) RF Company/Alamy Images. (r5–l) O. Louis Mazzatenta/National Geographic Image Collection. (r5–r) Iain McGillivray/Shutterstock. **60–61** (bg) Eastcott Momatiuk/Photodisc/Getty Images. **61** (t) Doug Sokell/Visuals Unlimited. (c) Edward Kinsman/Photo Researchers, Inc. (r) James L. Amos/National Geographic Image Collection. **62** (tl) Morozova Tatyana (Manamana)/Shutterstock. (tr) Mirka Moksha/Shutterstock. (cl) Harry Taylor/Dorling Kindersley/Getty Images. (c) Joyce Photographics/Photo Researchers, Inc. (cr) Milena Katzer/Shutterstock. (b) Linda/Shutterstock. **63** Jim Wark/Airphoto. **64** (l, r) RF Company/Alamy Images. (cl) Mark Schneider/Visuals Unlimited. (c) José Manuel Sanchis Calvete/Corbis. (cr) Greg C Grace/Alamy Images. **65** (l) Visuals Unlimited/Corbis. (cl) RF Company/Alamy Images. (c) Dawn Hagan/iStockphoto. (cr) Melissa Carroll/iStockphoto. (r) Biophoto Associates/Photo Researchers, Inc. **66** (t) Yoav Levy/Phototake/Alamy Images. (c) Scientifica/Visuals Unlimited. (r) DEA/A. Rizzi/De Agostini Picture Library/Getty Images. (c) Charles D. Winters/Photo Researchers, Inc. (cr) DEA/R. Appiani/De Agostini/Getty Images. (b) Ismael Montero Verdu/Shutterstock. **68** (l) david martyn hughes/Alamy Images. (r) Dirk Wiersma/Photo Researchers, Inc. **68–69** (bg) syringa/Shutterstock. **69** (t) PhotoDisc/Getty Images. **70** (l) Jim Sugar/Corbis. (r) PhotoDisc/Getty Images. **70–71** (bg) John Eascott and Yva Momatiuk/National Geographic Image Collection. **72** (l) Inge Johnsson/Alamy Images. **72–73** (bg) Tim Fitzharris/Minden Pictures/National Geographic Image Collection. **73** (l) Mike Brake/Shutterstock. **74** Sam Abell/National Geographic Image Collection. **75** (l, r) Andrew Northrup. (r) Todd Gipstein/National Geographic Image Collection. **76** (tl) derekfsmith/Shutterstock. (tr) Dorling Kindersley/Getty Images. (tr) Larry Stepanowicz/Visuals Unlimited. (c) Alessandro Colle/Alamy Images. **77** (t) Creatas/Jupiterimages. (tc) Gary Ombler/Dorling Kindersley/Getty Images. (tr) Mick Rock/Alamy Images. (bl) John Foxx Images/Imagestate. (bc) A.B. Joyce/Photo Researchers, Inc. (br) Andrew J. Martinez/Photo Researchers, Inc. **78** (t) ryasick photography/Shutterstock. (b) Juda Ngwenya/Reuters/Corbis. **79** (t) A. T. Willett/Alamy Images. (c) Eric Nathan/Alamy Images. (r) STR/epa/Corbis. **80** (t) PhotoDisc/Getty Images. (c) Mike Brake/Shutterstock. (r) BIOS Bios - Auteurs Delobelle Jean-Philippe/Peter Arnold, Inc. **80–81** (bg) Jim Sugar/Corbis. **81** (l) John Foxx Images/Imagestate. (c) A.B. Joyce/Photo Researchers, Inc. (r) Andrew J. Martinez/Photo Researchers, Inc. **82** 2009 Edward Fleming. **83** (l,r) Edward Fleming. **84–85** (bg) Tim Fitzharris/Minden Pictures/National Geographic Image Collection. **85** (t) Ralph Lee Hopkins/National Geographic Image Collection. (b) David Noble Photography/Alamy Images. **86** (t) Tom Bean/Corbis. (b) Tom Bean. **87** W.E. Garrett/National Geographic Image Collection. **88** (l) Tom Brownold Photography. (c) age fotostock. (r) Ralph Lee Hopkins/National Geographic Image Collection. **89** William H. Mullins/Photo Researchers, Inc **90** (t) Tom Bean. (b) William H. Mullins. **91** (t) Glenn Randall. (b) Brian Green/Alamy Images. **92** Tom Bean.

Chapter 3 **93, 94–95** Glen Allison/Photodisc/Getty Images. **96** (t) szefei/Shutterstock. (c) Bob Rowan; Progressive Image/Corbis. (b) James P. Blair/National Geographic Image Collection. **97** (t) Edmond Van Hoorick/Photodisc/Getty Images. (b) Richard Hutchings/PhotoEdit. **98** (t) szefei/Shutterstock. **99** (t) Dean Conger/National Geographic Image Collection. (tc) Vadim Ponomarenko/Shutterstock. (bc) Raymond Gehman/National Geographic Image Collection. (b) George F. Mobley/National Geographic Image Collection. **100** (t) Digital Vision/Getty Images. (b) image100/Alamy Images. **100–101** (bg) Alexei Fateev/Alamy Images. **101** (t) Derek Croucher/Alamy Images. **102** (t) Bob Rowan; Progressive Image/Corbis. **102–103** (bg) David R. Frazier Photolibrary, Inc./Alamy Images. **103** (b) PhotoDisc/Getty Images. **104–105** (bg) slavcic/Shutterstock. **104** (l, r) Andrew Northrup. **106–107** (bg) Henning de Beer/Gallo Images/Getty Images. **107** (l) Nick Norman/National Geographic Image Collection. (r) Dex Image/Alamy Images. **108** (l) Lee Prince/Shutterstock. (r) James P. Blair/National Geographic Image Collection. **108–109** (bg) Edmond Van Hoorick/Photodisc/Getty Images. **109** (t) Mikael Damkier/Shutterstock. (c) DigitalStock/Corbis. (bl) Maximilian Stock Ltd/Phototake/Alamy Images. (br) Jeffrey Coolidge/The Image Bank/Getty Images. **110–111** (bg) James P. Blair/National Geographic Image Collection. **111** (t) Steven Weinberg/Riser/Getty Images. (c) Claudius/Corbis. (b) Peter Bowater/Alamy Images. **112–113** Ilene MacDonald/Alamy Images. **114–115** Paul Roux/iStockphoto. **116–117** (bg) Tim Graham/Alamy Images. **117** (t) Wolfgang Hoffmann/AGStockUSA/Alamy Images. (b) NSIL/G.R. 'Dick' Roberts/Visuals Unlimited. (b) Maximilian Stock LTD/Phototake/Alamy Images. **118** (c) Wave Royalty Free/Photo Researchers, Inc. **118–119** (bg) J.D. Pooley/AP Images. **119** (t) PhotoDisc/Getty Images. **120–121** (bg) Nigel Lloyd/Alamy Images. **121** (r) Richard Hutchings/PhotoEdit. **122–123** (bg) Photoshot Holdings Ltd/Alamy Images. **123** (b) Brand X Pictures/Getty Images. **124** (cl) Colin Monteath/Minden Pictures/National Geographic Image Collection. **124–125** (bg) Sarah Leen/National Geographic Image Collection. **125** (l) Can Balcioglu/Shutterstock. (r) Fancy/Alamy Images. **126** (t) Papilio/Alamy Images. (b) Michael Nichols/National Geographic Image Collection. **126–127** (bg) Paul Zahl/National Geographic Image Collection. **127** (l, r) Mattias Klum/National Geographic Image Collection. **128** (bg) Joel Sartore/National Geographic Image Collection. (l) Bob Rowan; Progressive Image/Corbis. (r) Christina Richards/Shutterstock. **129** (b) Image Source/Getty Images. **130** (t) PhotoDisc/Getty Images. (b) Johari Cole. **131** (t) PhotoDisc/Getty Images. (b) Johari Cole. **132** Photodisc/Getty Images. **133** inga spence/Alamy Images. **134** (tl) James Strawser/Grant Heilman Photography. (tr) Alan Pitcairn/Grant Heilman Photography. (c) Peter Pattavina/iStockphoto. (b) Simon Rawles/Alamy Images. **135** (t) Terry Brandt/Grant Heilman Photography. (b) Michael S. Yamashita/Corbis. **136** Walter Hodges/Photographer's Choice RF/Getty Images. **137** (t) Michael Rosenfeld/Science Faction/Corbis. (b) Joerg Boethling/Alamy Images. **138** (t) Janis Christie/Photodisc/Getty Images. (b) Steven Weinberg/Stone/Getty Images. **139** Steve Hix/Somos Images/Corbis. **140** Alan Pitcairn/Grant Heilman Photography.

Chapter 4 **141, 142–143** Jonas Bendiksen/National Geographic Image Collection. **144** (t) Kazuyoshi Nomachi/Corbis. (c) Lee Foster/Lonely Planet Images/Getty Images. **145** (t) Ludo Kuipers/Corbis. (b) Gerald & Buff Corsi/Visuals Unlimited. **146–147** (bg) Kazuyoshi Nomachi/Corbis. **147** (t) James L. Stanfield/National Geographic Image Collection. (tc) James Steinberg/Photo Researchers, Inc. (tb) Momatiuk Eastcott/Corbis. (bt) P.Michael Photoz/AKA/age fotostock. (bc) Wilfried Krecichwost/Digital Vision/Getty Images. (b) David Noble Photography/Alamy Images. **148–149** (bg) Lee Foster/Lonely Planet Images/Getty Images. **149** (inset) Kazuyoshi Nomachi/Corbis. **150–151** (bg) Patrick Ward/Corbis. **151** (inset) David Boyer/National Geographic Image Collection. **152–153** (b) Ludo Kuipers/Corbis. **153** (tl, tr) Andrew Northrup. **154** (c) Corbis. **154–155** (bg) Ron Watts/Corbis. **156** (c) Ron Adcock/Alamy Images. **156–157** (bg) Fritz Polking/Peter Arnold, Inc. **158–159** James Randklev/Corbis. **160** (t) Yannis Samatas. **160–161** (b) Atlantide Phototravel/Corbis. **161** (c) NASA Goddard Space Flight Center. **162** (c) Michael Nichols/National Geographic Image Collection. **162–163** (bg) Jason Edwards/National Geographic Image Collection. **164** (t) Dean Conger/Corbis. **164–165** (bg) Gerald & Buff Corsi/Visuals Unlimited. **165** (t) David Paterson/Photographer's Choice/Getty Images. (c) AP Images. **166–167** (bg) Bullit Marquez/AP Images. **167** (inset) Mike Eliason/Santa Barbara News-Press/Zuma Press. **168** (t) LeveretBradley/Corbis Premium RF/Alamy Images. **168–169** (b) FLariviere/Shutterstock. **169** (c) Steven Frame/Shutterstock. **170** (c) Richard Olsenius/National Geographic Image Collection. **170–171** (bg) Eye Ubiquitous/Alamy Images. **171** (c) DEA/G. Dagli Orti/De Agostini/Getty Images. **172** (t) Neil C. Robinson/Corbis. (c, b) James Brunker/Alamy Images. **173** (tl) Martin Alipaz/Corbis. (tc) Arco Images/Philips R/age fotostock. (tr) Annie Belt/Corbis. (b) Martin Alipaz/EFE/epa/Corbis. **174** (l) DigitalStock/Corbis. (c) Panoramic Images/Getty Images. **174–175** (bg) World Perspectives/Stone/Getty Images. **175** (l) LeighSmithImages/Alamy Images. **176–177** (l) Dr. Lonnie G. Thompson. **178–179** (b) Creatas/Jupiterimages. **179** (t) Image Source/Getty Images. (c) Pali Arts Communications/National Geographic Image Collection. **180** (l) Courtesy of the National Park Service, Yosemite National Park, YOSE 21377. (c) Courtesy of the National Park Service, Yosemite National Park, YOSE 21380. (r) Courtesy of the National Park Service, Yosemite National Park, YOSE 21381. **181** Krys Bailey/Alamy Images. **182** Alex Neauville/Shutterstock. **183** Corbis RF/Alamy Images. **184** Carr Clifton/Minden Pictures/National Geographic Image Collection. **186–187** David Muench/Corbis. **188** Creatas/Jupiterimages.

Chapter 5 **189, 190–191** Steve and Donna O'Meara/National Geographic Image Collection. **192** (t) Ralph Lee Hopkins/National Geographic Image Collection. (c) Roger Ressmeyer/Corbis. (b) Oded Balilty/AP Images. **193** (c) Bryan Lowry/SeaPics.com. (b) J. Baylor Roberts/National Geographic Image Collection. **194–195** D Woodfall/ImageState/Panoramic Images. **195** (l) Ralph Lee Hopkins/National Geographic Image Collection. (r) James P. Blair/National Geographic Image Collection. **196–197** (bg) Roger Ressmeyer/Corbis. **197** (l, r) Andrew Northrup. **198** (inset) Achmad Ibrahim/AP Images. **198–199** (bg) Oded Balilty/AP Images. **199** (inset) Roger Ressmeyer/Corbis. **200–201** (bg) Emory Kristof/National Geographic Image Collection. **201** (t) Ewing Krainin/Photo Researchers, Inc. (c) Michael T. Sedam/Corbis Premium RF/Alamy Images. (b) J. Baylor Roberts/National Geographic Image Collection. **202** (inset) Emmanuel Lattes/Alamy Images. **202–203** (bg) Alberto Garcia. **203** (inset) Ho/Reuters/Corbis. **204** (inset) Data courtesy Landsat 7 project and EROS Data Center. Caption by James Foster, NASA Goddard Space Flight Center/NASA - Earth Observatory. **204–205** (bg) D.M. Peterson, U.S. Geological Survey. **206** (t) Corbis Super RF/Alamy Images. (b) Roger Ressmeyer/Corbis. **207** (t) Itsuo Inouye/AP Images. (b) Kyodo via/AP Images. **208** (t) Adam DuBrowa/FEMA. (c) Emory Kristof/National Geographic Image Collection. (r) D.M. Peterson, U.S. Geological Survey. **208–209** (bg) UPI Photo/Carlos Gutierrez/NewsCom. **210–211** David Pyle. **212–213** (bg) Michael T. Sedam/Corbis. **213** (inset) Jacques Descloitres, MODIS Land Rapid Response Team at NASA GSFC. **216** Debbie Yea/Alamy Images. (b) Amos Zezmer/Orange Stock/age fotostock. **217** (t) dave jepson/Alamy Images. (b) Bryan Lowry/SeaPics.com. **218–219** Roger Ressmeyer/Corbis. **220** Dave Jepson/Alamy Images.

Chapter 6 **221, 222–223** Digital Vision/Getty Images. **224** (t) Science Photo Library/Photo Researchers, Inc. (c) image100/Photolibrary. **225** (t) Jason Edwards/National Geographic Image Collection. (c) Earth Imaging/Stone/Getty Images. (b) kavram/Shutterstock. **226–227** Science Photo Library/Photo Researchers, Inc. **227** (bl, br) Andrew Northrup. **228–229** NASA - digital version copyright/Science Faction/Corbis. **230–231** Paul Burns/Blend Images/Photolibrary. **231** (tl) Judith Collins/Alamy Images. (tc) Alan & L. Detrick/Photo Researchers, Inc. (tr) Jaywarren79/Shutterstock. (bl) Duncan Walker/iStockphoto. (bc) David J. Green - technology/Alamy Images. (br) Donall O Cleirigh/iStockphoto. **232–233** Jose Luis Pelaez, Inc./Corbis. **233** (t) BrandX/Jupiterimages. (ct) Dennis MacDonald/PhotoEdit. (c) Gabe Palmer/Alamy Images. (cb) Dirk v. Mallinckrodt/Alamy Images. (b) PhotoDisc/Getty Images. **234** (b) Corbis Super RF/Alamy Images. **234–235** Corbis Super RF/Alamy Images. **235** (inset) Comstock Images/Photolibrary. **236–237** Tanya Constantine/Blend Images/Getty Images. **238** (l) Lars Christensen/Shutterstock. **238–239** (bg) Jason Edwards/National Geographic Image Collection. (c) paul prescott/Shutterstock. **239** (r) Jim Harris/Shutterstock. **240** (t) John Eascott and Yva Momatiuk/National Geographic Image Collection. **240–241** Corbis/Photolibrary. **242–243** (bg) image100/Photolibrary. **243** (inset) Cranes in the Rain/Shutterstock. **244–245** Corbis/age fotostock. **245** (b) Martin Maun/Shutterstock. **246–247** Roca/Shutterstock. **248–249** kavram/Shutterstock. **250–251** David Sutherland/Photographer's Choice/Getty Images. **254** (b) Matthew Ragen/iStockphoto. **254–255** (bg) DigitalStock/Corbis. **256** (cl) John Eascott and Yva Momatiuk/National Geographic Image Collection. (b) Duncan Walker/iStockphoto. **256–257** kavram/Shutterstock. **257** (inoct) Colin Hawkins/Taxi/Getty Images. **258** Veronica Johnson. **259** (l) NOAA/Getty Images. (r) Veronica Johnson. **260–261** Yoann Combronde/Shutterstock. **261** (tl) Elnur/Shutterstock. (b) Tom Smith. **262** Mitsuaki Iwago/Minden Pictures/National Geographic Image Collection. **263** (cr) Wave Royalty Free/Alamy Images. **264** (t) The Pilot's-Eye View/Shutterstock. (b) Jay Dickman/National Geographic Image Collection. **265** Carsten Peter/National Geographic Image Collection. **266–267** Kenneth Garrett/National Geographic Image Collection. **267** Ed Darack/Science Faction/Corbis. **268** Wave Royalty Free/Alamy Images.

End Matter **EM1** Ludo Kuipers/Corbis. **EM2** Richard Hutchings/PhotoEdit. **EM7** Atlantide Phototravel/Corbis. **Back Cover** (bg) Kevin M. Law/Alamy Images. (tl) Beverly Goodman. (tr) Amir Yurman/The Leon Recanati Institute for Maritime Studies. (c) Hanan Isachar/JAI/Corbis. (bl) InterNetwork Media/Digital Vision/Getty Images. (bl) Radius Images/Corbis.